METHODS IN MOLECULAR BIOLOGY

Series Editor
John M. Walker
School of Life and Medical Sciences
University of Hertfordshire
Hatfield, Hertfordshire, AL10 9AB, UK

For further volumes:
http://www.springer.com/series/7651

Mouse Embryogenesis

Methods and Protocols

Edited by

Paul Delgado-Olguin

Translational Medicine, The Hospital for Sick Children, Toronto, ON, Canada; Department of Molecular Genetics, University of Toronto, Toronto, ON, Canada; Heart & Stroke/Richard Lewar Centres of Excellence in Cardiovascular Research, Toronto, ON, Canada

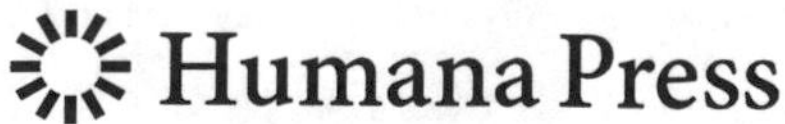

Editor
Paul Delgado-Olguin
Translational Medicine
The Hospital for Sick Children
Toronto, ON, Canada

Department of Molecular Genetics
University of Toronto
Toronto, ON, Canada

Heart & Stroke/Richard Lewar Centres of Excellence
in Cardiovascular Research
Toronto, ON, Canada

ISSN 1064-3745 ISSN 1940-6029 (electronic)
Methods in Molecular Biology
ISBN 978-1-4939-9265-2 ISBN 978-1-4939-7714-7 (eBook)
https://doi.org/10.1007/978-1-4939-7714-7

Softcover re-print of the Hardcover 1st edition 2018

Printed on acid-free paper

This Humana Press imprint is published by the registered company Springer Science+Business Media, LLC part of Springer Nature.
The registered company address is: 233 Spring Street, New York, NY 10013, U.S.A.

Preface

The mouse has been instrumental in our quest to understand the processes that control embryogenesis. The mouse reaches sexual maturity at 2 months of age, and after 21 days of gestation it produces litters of 8–20 pups. The life cycle of mice and its adaptability to life in specialized research facilities have made it an ideal model to study mammalian development. In addition, high similarity between the mouse and human genomes has made the mouse instrumental in the discovery of fundamental mechanisms controlling human development. With the advent of site-directed mutagenesis in the mouse, we can inactivate every single gene in specific cell types at specific time points to interrogate gene function in developmental processes. This has allowed the identification and dissection of pathways controlling not only embryonic development but also postnatal organ function and maintenance. Targeted gene mutation also revealed that mutation of ortholog genes in the mouse recapitulates human disease.

This volume of the *Methods in Molecular Biology* series provides a collection of protocols for identification of mutant mice and characterizing aspects of their anatomical, functional, cellular, and molecular phenotypes. Each chapter opens with a description of the fundamentals of the protocol to be described. The Materials section lists the chemicals and reagents and provides instructions for preparing the solutions required to perform the protocol. Detailed instructions for every step of the protocol will ensure its successful execution by the experimentalist. Each chapter also provides notes, sharing tips to facilitate procedure through specific steps of the protocol, and that will guide the experimentalist through troubleshooting. The protocols in this book are varied in nature and treat with identification of mutant mice, anatomical and functional phenotyping using quantitative imaging, isolation of specific embryonic cell types for cell culture or molecular analysis, analysis of gene expression, and characterization of chromatin structure. In addition, two chapters provide an overview of novel approaches to induce directed mutagenesis and to characterize genome dynamics. These approaches are already broadening our view of the mechanisms controlling embryogenesis. The advent of genomics approaches and the discovery of novel gene regulators leave us now with limitless opportunities to discover new aspects of embryogenesis control, organ function, and the origins of disease. It is my hope that this book helps experimentalists in this endeavor.

Toronto, ON, Canada *Paul Delgado-Olguin*

Contents

Contributors

ABDALLA AHMED · *Translational Medicine, The Hospital for Sick Children, Peter Gilgan Centre for Research and Learning, Toronto, ON, Canada; Department of Molecular Genetics, University of Toronto, Toronto, ON, Canada*

RODRIGO G. ARZATE-MEJÍA · *Departamento de Genética Molecular, Instituto de Fisiología Celular, Universidad Nacional Autónoma de México, Mexico City, Mexico*

CLAUDIA BILODEAU · *Program in Translational Medicine, Hospital for Sick Children, Peter Gilgan Centre for Research and Learning, Toronto, ON, Canada; Department of Laboratory Medicine and Pathobiology, University of Toronto, Toronto, ON, Canada*

LINDSAY S. CAHILL · *The Hospital for Sick Children, Mouse Imaging Centre, Toronto, ON, Canada*

YONG CHEN · *Translational Medicine Program, Division of General and Thoracic Surgery, The Hospital for Sick Children, Toronto, ON, Canada*

LIJUN CHI · *Translational Medicine, The Hospital for Sick Children, Peter Gilgan Centre for Research and Learning, Toronto, ON, Canada; Department of Molecular Genetics, University of Toronto, Toronto, ON, Canada*

PAUL DELGADO-OLGUIN · *Translational Medicine, The Hospital for Sick Children, Peter Gilgan Centre for Research and Learning, Toronto, ON, Canada; Department of Molecular Genetics, University of Toronto, Toronto, ON, Canada; Heart & Stroke/ Richard Lewar Centres of Excellence in Cardiovascular Research, Toronto, ON, Canada*

MAYRA FURLAN-MAGARIL · *Departamento de Genética Molecular, Instituto de Fisiología Celular, Universidad Nacional Autónoma de México, Mexico City, Mexico*

ALEJANDRA GARATE-CARRILLO · *Department of Medicine, School of Medicine, University of California, San Diego, CA, USA*

ESTELA GARCÍA-GONZÁLEZ · *Sociedad Mexicana de Epigenética y Medicina Regenerativa, A.C., Coyoacán, Ciudad de México, Mexico*

J. MANUEL HERNÁNDEZ-HERNÁNDEZ · *Unidad de Investigación Médica en Genética Humana, Hospital de Pediatría, Colonia Doctores, Ciudad de México, Mexico; Sociedad Mexicana de Epigenética y Medicina Regenerativa, A.C., Coyoacán, Ciudad de México, Mexico*

LISA L. HUA · *Cardiovascular Research Institute, University of California, San Francisco, CA, USA*

MIKIHIRO INOUE · *Department of Gastrointestinal and Pediatric Surgery, Mie University Graduate School of Medicine, Tsu, Mie, Japan*

KARINA JÁCOME-LÓPEZ · *Departamento de Genética Molecular, Instituto de Fisiología Celular, Universidad Nacional Autónoma de México, Mexico City, Mexico*

YUHKI KOIKE · *Translational Medicine Program, Division of General and Thoracic Surgery, The Hospital for Sick Children, Toronto, ON, Canada; Department of Gastrointestinal and Pediatric Surgery, Mie University Graduate School of Medicine, Tsu, Mie, Japan*

HISATO KONOEDA · *Translational Medicine, The Hospital for Sick Children, Toronto, ON, Canada*

KAZUKO KOSHIBA-TAKEUCHI · *Faculty of Life Sciences, Toyo University, Ora-gun, Gunma, Japan*

MASATO KUSUNOKI · *Department of Gastrointestinal and Pediatric Surgery, Mie University Graduate School of Medicine, Tsu, Mie, Japan*
IRINA V. LARINA · *Department of Molecular Physiology and Biophysics, Baylor College of Medicine, Houston, TX, USA*
CAROL LEE · *Translational Medicine Program, Division of General and Thoracic Surgery, The Hospital for Sick Children, Toronto, ON, Canada*
JU HEE LEE · *Translational Medicine, The Hospital for Sick Children, Toronto, ON, Canada; Department of Laboratory Medicine and Pathobiology, University of Toronto, Toronto, ON, Canada*
BO LI · *Translational Medicine Program, Division of General and Thoracic Surgery, The Hospital for Sick Children, Toronto, ON, Canada*
PAULA LICONA-LIMÓN · *Departamento de Biología Celular y del Desarrollo, Instituto de Fisiología Celular, Universidad Nacional Autónoma de México, Mexico City, Mexico*
ANDREW L. LOPEZ III · *Department of Molecular Physiology and Biophysics, Baylor College of Medicine, Houston, TX, USA*
TAKASHI MIKAWA · *Department of Anatomy, Cardiovascular Research Institute, University of California, San Francisco, CA, USA*
HIROMU MIYAKE · *Translational Medicine Program, Division of General and Thoracic Surgery, The Hospital for Sick Children, Toronto, ON, Canada*
JOON HO MOON · *Translational Medicine, The Hospital for Sick Children, Toronto, ON, Canada*
CARLOS PALMA-FLORES · *Sociedad Mexicana de Epigenética y Medicina Regenerativa, A.C., Coyoacán, Ciudad de México, Mexico; Catedrático CONACYT-Instituto Politécnico Nacional, Ciudad de México, Mexico*
AGOSTINO PIERRO · *Translational Medicine Program, Division of General and Thoracic Surgery, The Hospital for Sick Children, Toronto, ON, Canada; The Hospital for Sick Children, University of Toronto, Toronto, ON, Canada*
MARTIN POST · *Program in Translational Medicine, Hospital for Sick Children, Peter Gilgan Centre for Research and Learning, Toronto, ON, Canada; Department of Laboratory Medicine and Pathobiology, University of Toronto, Toronto, ON, Canada*
ISRAEL RAMIREZ · *Universidad Panamericana, Escuela de Medicina, Ciudad de México, Mexico*
FÉLIX RECILLAS-TARGA · *Departamento de Genética Molecular, Instituto de Fisiología Celular, Universidad Nacional Autónoma de México, Mexico City, Mexico*
BLADIMIR ROQUE-RAMIREZ · *Sociedad Mexicana de Epigenética y Medicina Regenerativa, A.C., Coyoacán, Ciudad de México, Mexico*
ANNA R. ROY · *Translational Medicine, The Hospital for Sick Children, Peter Gilgan Centre for Research and Learning, Toronto, ON, Canada; Department of Molecular Genetics, University of Toronto, Toronto, ON, Canada; Genomics and Genome Biology, The Hospital for Sick Children, Peter Gilgan Centre for Research and Learning, Toronto, ON, Canada*
JOHN G. SLED · *The Hospital for Sick Children, Mouse Imaging Centre, Toronto, ON, Canada; Department of Medical Biophysics, University of Toronto, Toronto, ON, Canada*
HOON-KI SUNG · *Translational Medicine, The Hospital for Sick Children, Toronto, ON, Canada; Department of Laboratory Medicine and Pathobiology, University of Toronto, Toronto, ON, Canada*
KEIICHI UCHIDA · *Department of Gastrointestinal and Pediatric Surgery, Mie University Graduate School of Medicine, Tsu, Mie, Japan*

SANDRA VUONG · *Translational Medicine, The Hospital for Sick Children, Peter Gilgan Centre for Research and Learning, Toronto, ON, Canada; Department of Molecular Genetics, University of Toronto, Toronto, ON, Canada*

SHANG WANG · *Department of Molecular Physiology and Biophysics, Baylor College of Medicine, Houston, TX, USA*

RICHARD WU · *Department of Gastroenterology, Hepatology and Nutrition, The Hospital for Sick Children, Toronto, ON, Canada*

AZADEH YEGANEH · *Translational Medicine, The Hospital for Sick Children, Toronto, ON, Canada*

BEHZAD YEGANEH · *Program in Translational Medicine, Hospital for Sick Children, Peter Gilgan Centre for Research and Learning, Toronto, ON, Canada*

YU-QING ZHOU · *The Hospital for Sick Children, Mouse Imaging Centre, Toronto, ON, Canada; Ted Rogers Centre for Heart Research, Translational Biology & Engineering Program, Institute of Biomaterial & Biomedical Engineering, University of Toronto, Toronto, ON, Canada*

Chapter 1

Mouse Genotyping

Sandra Vuong and Paul Delgado-Olguin

Abstract

Genotyping is an invaluable tool for identifying organisms carrying genetic variations including insertions and deletions. This method involves the extraction of DNA from animal tissue samples and subsequent amplification of genomic regions of interest by polymerase chain reaction (PCR). The amplified products are analyzed by agarose gel electrophoresis based on their size.

Key words Genotype, DNA, Polymerase chain reaction

1 Introduction

There are about 24,000 protein-coding genes in the mouse genome, and the function of many of them has yet to be elucidated [1]. In a realm where genetically modified mouse lines are instrumental to discovering the function of unknown genes, identifying mice carrying genetic modification (s) affecting the function of specific genes is required.

Genotyping mice involves the amplification of a genomic region of interest from isolated DNA. The DNA is commonly extracted from a small piece of tissue, most commonly from a tail clip, ear notch, or the yolk sac. Primers designed to target the genomic region or gene of interest are used for amplification by polymerase chain reaction (PCR) [2–4]. Primers can be designed using many freely available software platforms such as PrimerBLAST [5, 6]. The PCR amplified products are then separated by gel electrophoresis and visualized using a gel documentation system. Genotypes are then determined by identifying PCR products of the corresponding expected size(s) determined by the location of the primers used.

Two methods of DNA isolation are described below. The quick method is simpler, as it does not require elimination of excess protein. Thus, amplification with some primers may require optimization. However, once optimized, this method consistently

Paul Delgado-Olguin (ed.), *Mouse Embryogenesis: Methods and Protocols*, Methods in Molecular Biology, vol. 1752,
https://doi.org/10.1007/978-1-4939-7714-7_1, © Springer Science+Business Media, LLC, part of Springer Nature 2018

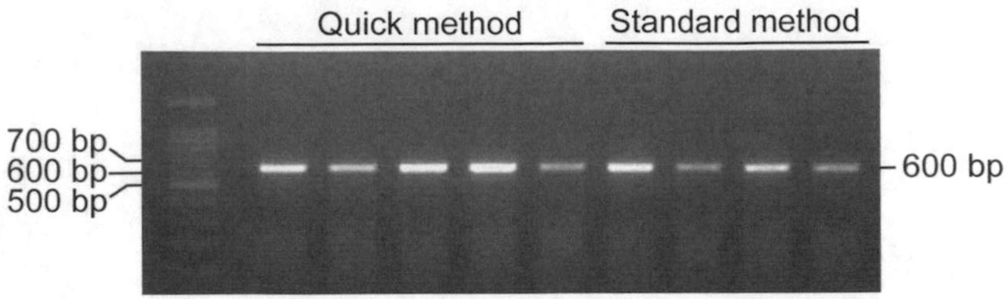

Fig. 1 Genotyping of samples using both the quick and standard DNA extraction methods using the same primer pair. Using DNA isolated through both the quick or standard methods as template in PCR produces the expected 600 bp fragment

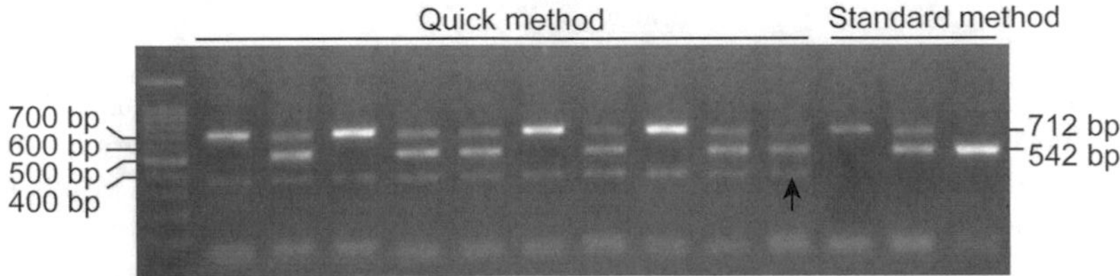

Fig. 2 Differences in genotyping of samples using two different DNA extraction methods. The arrow indicates an unspecific band amplified from DNA isolated through the quick method. Amplification using DNA isolated through the standard method as template did not produce the unspecific band. Both methods were useful to amplify the wild type (542 bp) and the mutant (712 bp) PCR products

produces specific PCR products. The standard method requires incubation with Proteinase K to eliminate excess protein, and thus takes a longer time to perform. However, this method often requires less optimization than the quick method. In our experience, the quick method works very well for samples obtained from mouse embryos and yolk sacs (which are easier to digest than tail clips and ear notches). For templates difficult to amplify, the standard method often works better. The results from each method are compared in Figs. 1 and 2.

2 Materials

All solutions should be prepared using double distilled water (MilliQ water). Solutions can be prepared in advance and stored at room temperature unless otherwise stated.

2.1 Quick DNA Isolation

1. *50 mM NaOH*: Add 250 μL of 10 N sodium hydroxide to 49.75 mL of water for a total of 50 mL.
2. *0.5 M Tris–HCl pH 8.0*: To prepare 1 M Tris–HCl pH 8.0 stock solution, weigh 60.55 g of Tris base and dissolve in 400 mL of water. Adjust the pH to 8.0 by adding concentrated hydrochloric acid, 37% in solution. Allow solution to acclimate to room temperature before bringing the volume of the

solution to 500 mL with water. Add 25 mL of water to 25 mL of 1 M Tris–HCl solution to make 50 mL of 0.5 M Tris–HCl pH 8.0.

2.2 Standard DNA Isolation

1. *DNA isolation buffer*: 200 mM NaCl, 100 mM Tris-HCl pH 8.5, 5 mM EDTA pH 8.0, 0.5% Tween 20. To prepare 5 M NaCl stock solution, dissolve 292.44 g of sodium chloride in 1 L of water. To prepare 1 M Tris–HCl stock solution weigh 60.55 g of Tris base and dissolve in 400 mL of water. Adjust the pH of the solution to 8.5 by addition of concentrated hydrochloric acid, 37% in solution. Let the solution acclimate to room temperature prior to bringing the volume up to 1 L with water. To prepare 0.5 M EDTA stock solution dissolve 93.05 g of EDTA in 400 mL of water. Adjust the pH to 8.0 by adding concentrated 10 N NaOH. Let the solution adjust to room temperature before bringing the final volume to 1 L with water. Autoclave the 0.5 M EDTA solution for 20 min on liquid cycle.

   To make 500 mL solution of DNA isolation buffer, add 20 mL of 5 M NaCl, 50 mL of 1 M Tris pH 8.5, 5 mL of 0.5 M EDTA, and 2.5 mL of Tween 20 to 447.5 mL of water.
2. *Proteinase K solution*: 10 mg/mL. Weigh 200 mg of Proteinase K and dissolve in 20 mL of water. Prepare aliquots of 1 mL in 1.5 mL microcentrifuge tubes for storage. Store aliquots at −20 °C.

2.3 Polymerase Chain Reaction (PCR)

1. *10× PCR Buffer*: 100 mM Tris pH 8.3, 500 mM KCl, 20 mM $MgCl_2$. To prepare 1 M Tris pH 8.3 stock solution, dissolve 60.55 g of Tris base in 400 mL of water. Adjust pH to 8.3 by slowly adding concentrated HCl, 37% in solution. Bring solution to room temperature before adjusting final volume to 500 mL with water. To make 1 M KCl stock solution, dissolve 37.275 g of potassium chloride in 400 mL of water. Bring the final volume up to 500 mL and autoclave for 20 min on liquid cycle. To prepare 1 M $MgCl_2$ stock solution, weigh 101.65 g of magnesium chloride hexahydrate and dissolve in 400 mL of water. Bring final volume to 500 mL. Autoclave for 20 min on liquid cycle.

   To make a 20 mL solution of 10× PCR Buffer, add 2 mL of 1 M Tris pH 8.3, 10 mL 1 M KCl, and 2.4 mL 1 M $MgCl_2$ to 5.6 mL of water. Aliquot 1 mL of 10× PCR buffer into 20 1.5 mL microcentrifuge tubes for storage. Store at −20 °C (*see* **Note 1**).
2. *2 mM dNTPs*: Combine 250 μL of 100 mM dATP, dCTP, dGTP, and dTTP in 11.5 mL of water for a total volume of 12.5 mL. Aliquot 250 μL of 2 mM dNTPs into 50 1.5 mL microcentrifuge tubes for storage. Store at −20 °C (*see* **Note 2**).

3. *5 μM primer pair solution*: First, make 100 μM primer stock solutions of both forward and reverse oligonucleotides. To make 100 μL of 5 μM working primer pair solution, add 10 μL of forward primer and 10 μL of reverse primer to 80 μL of water. Store at −20 °C.
4. *Taq polymerase*: Taq polymerase may be commercially purchased (*see* **Note 3**).
5. *6× Orange G loading dye:* 0.02% Orange G (weight/volume) in 30% sucrose. Sucrose solution can be prepared by dissolving 15 g of sucrose in water. To make loading dye, weigh 2 mg of Orange G dye and dissolve in 10 mL of 30% sucrose (*see* **Note 4**).

2.4 Gel Electrophoresis

1. *TAE buffer*: Prepare 1 L of 50× TAE Buffer stock solution by dissolving 24.2 g of Tris Base in 700 mL water. Add 100 mL of 0.5 M EDTA pH 8.0 and 57.1 mL of Glacial Acetic Acid to the solution. Bring the solution to a final volume of 1 L. To make 1× TAE Buffer, add 20 mL of 50× TAE solution to 980 mL of water.
2. *Agarose.*
3. *Nucleic acid gel staining solution* (*see* **Note 5**).
4. *DNA ladder* (*see* **Note 6**).

3 Methods

All steps should be carried out at room temperature unless otherwise stated. The quick method is recommended for embryonic tissues. The standard method often works better with DNA isolated from ear notches and tail clips.

3.1 Sample Collection

1. An embryonic yolk sac, or a piece of embryonic tale of 3 mm length is sufficient for DNA extraction and subsequent genotyping. Cut approximately 3 mm of tissue from the embryo's tail and place it into a sterile 1.5 mL microcentrifuge tube (*see* **Note** 7).

3.2 DNA Extraction

3.2.1 Quick Method

1. To each microcentrifuge tube, add 300 μL of 50 mM NaOH. Make sure that the sample is submerged into the solution (*see* **Note 8**).
2. Incubate samples at 95 °C for 15 min to disaggregate tissue (*see* **Note 9**).
3. Add 100 μL of 0.5 M Tris–HCl pH 8.0 to neutralize pH and stop the reaction. Mix well. Spin at 5k rmp for 1 min to bring down debris. The sample is now ready for PCR (*see* **Note 10**).

3.2.2 Standard Method

1. Add 400 μL of DNA isolation Buffer and 50 μL of 10 mg/mL Proteinase K to each sample. Mix. Make sure that the sample is submerged in the solution.
2. Incubate the sample at 55 °C overnight (*see* **Note 11**).
3. Incubate samples at 95 °C for 10 min to inactivate Proteinase K. Spin at 5k rmp for 1 min to bring down debris. Samples are now ready for PCR (*see* **Note 12**).

3.3 Polymerase Chain Reaction

1. Thaw 10× PCR buffer, 2 mM dNTPs and primer working solution(s).
2. In sterile PCR tube(s), set up each reaction (total volume 20 μL) as follows:

   2 μL 10× PCR Buffer, 2 μL 2 mM dNTPs, 2 μL primer mix, 0.5 μL *Taq* Polymerase, 11.5 μL water, and 2 μL DNA sample (*see* **Note 13**).
3. Set up a PCR program on a thermocycler as follows:
   (a) 3:00 min at 95 °C. This step is for initial denaturation of DNA.
   (b) 0:30 s at 95 °C. This step is for denaturation of DNA.
   (c) 0:30 s at specific annealing temperature for primers (*see* **Note 14**).
   (d) 1:00 min at 72 °C. This step allows for elongation of DNA strands (*see* **Note 15**).
   (e) Repeat **steps 2–4** for a total of 35 cycles. This step allows for exponential amplification of DNA template (*see* **Note 16**).
   (f) 10:00 min at 72 °C. Final extension of DNA templates occurs here.
   (g) 10:00 min at 4 °C. PCR products may then be removed and stored at 4 °C until analysis (*see* **Note 17**).

3.4 Gel Electrophoresis

1. Weigh enough agarose to prepare the gel at the required volume and percentage. The percentage of the gel should be chosen based on the DNA fragment size. To separate small DNA fragments a higher percentage is required. Measure the required amount of TAE buffer to cast the agarose gel. Mix the agarose and TAE buffer in a clean flask (*see* **Note 18**).
2. Microwave to heat the solution in order to fully dissolve the agarose in TAE buffer (*see* **Note 19**).
3. Let the solution cool down at room temperature. After a few minutes, add the nucleic acid gel stain to the mixture. Mix well. When the flask is cool enough to be touched by a gloved hand, it is ready for pouring into gel cast (*see* **Note 20**).
4. Let the gel polymerize at room temperature for a minimum of 20 min (*see* **Note 21**).

5. Remove the comb from the gel and place the gel into an electrophoresis tank. Fill the tank with enough TAE buffer to cover the gel. Load samples and the DNA ladder into the wells (*see* **Note 22**).
6. Let the samples run at 120 V for 30 min (*see* **Note 23**).
7. Image the gel using a gel documentation system.

4 Notes

1. 10× PCR reaction buffer may be purchased with *Taq Polymerase*.
2. 2 mM dNTPs may be replaced by the dNTP mix that is included with commercially purchased *Taq Polymerase*. Remember to take note of the concentration of commercial dNTP reagent mix and make adjustments to PCR recipe accordingly.
3. As an alternative to purchasing commercial *Taq* polymerase, the enzyme can be purified in the lab using a published protocol [7].
4. Orange G dye may be replaced with other dyes like bromophenol blue and xylene cyanol.
5. Commercially available nucleic acid gel staining solutions that have been designed as safer alternatives to ethidium bromide (EtBr) such as *RedSafe*, *GelRed*, and *SYBR Safe*, are recommended.
6. DNA ladders can be commercially purchased. The size of the DNA ladder that should be used depends on the PCR product size. For example, 100 bp DNA ladders are recommended for PCR templates that are shorter than 1500 bp. 1 kb DNA ladders should be used for templates that are a few kilobases long.
7. As an alternative to a tail clip, a hind limb or forelimb from embryos at embryonic day (*E*) 12.5 or older can be used. Yolk sac membranes may also be used. For pups that are younger than 21 days (P21), either a tail clip of approximately 5 mm, or a single ear notch is sufficient. For adult mice, ear notch samples should be collected. While fresh samples are ideal for extraction, samples may be also stored at −20 °C for future DNA extraction.
8. The amount of 50 mM NaOH added is adjusted based on sample size. For samples that are smaller than approximately 3 mm, 225 μL of 50 mM NaOH are added. For samples larger than 7 mm, the volume of 50 mM NaOH is increased to 450 μL. For ear notches, 300 μL 50 mM NaOH are added.

9. Incubation time can be extended based on the samples used. An incubation period of 10–15 min is ideal for fresh embryonic tissue. For ear notches, incubation of 20–30 min is best. The degree of tissue disaggregation can be periodically checked, and tubes manually agitated for better digestion. Samples may be removed from 95 °C when all the tissue has been disintegrated.
10. To quench the reaction, the amount of 0.5 M Tris–HCl pH 8.0 required is 1/3 the volume of NaOH used. Samples can be stored at 4 °C for short term or −20 °C for long term.
11. Alternatively, samples can be digested at 55 °C for approximately 5 h.
12. Samples may be stored at 4 °C for a couple of days, or at −20 °C for the long term.
13. If analyzing more than one sample, preparing a master mix with all components excluding DNA is recommended. Additionally, 4 μL of 6× loading dye can be added to each reaction at this step to make a total reaction volume of 24 μL. Water and *Taq* polymerase volumes can be adjusted according to manufacturer's instructions.
14. To determine optimal annealing temperature for a specific primer pair, running an initial gradient PCR is recommended. Usually a gradient PCR with temperatures ranging from 53 to 70 °C is sufficient to determine optimal primer annealing temperature. For PCR in Fig. 1 we used primers F: ATCCGAAAAGAAAACGTTGA, and R: ATCCAGGTTACGGATATAGT, which amplify a 600 bp of the gene encoding the Cre recombinase. For Fig. 2, we used primers F: AGGTTGTGAGCTGCCATATAA, and R: CTGTCGGAAAGGGTACTTCAT, which amplify two PCR products to distinguish between a wild type (542 bp), and a mutant allele (712 bp) containing a *loxP* site in the *L3mbtl2* gene [8].
15. An extension time of 1:00 min is sufficient for amplification of products expected to 1 kb in length. For longer fragments, extension time may be increased. As a general rule for calculation, use an extension time of 0:30 s for every 500 bp to be amplified.
16. The total number of cycles may be adjusted to increase or decrease final template concentration. 25–32 cycles are usually sufficient for amplification. Increasing the reaction over 35 cycles is not recommended, as the likelihood of obtaining unspecific bands increases.
17. PCR products may be stored at 4 °C for a period of up to 1 week before gel electrophoresis. Alternatively, PCR products may be stored at −20 °C for long term.

18. The percentage of agarose gel should be chosen based on the size of the DNA fragment that needs to be resolved. For example, 1.5% (weight/volume) agarose gels resolve fragments between 200 and 3000 bp. Keep in mind that the agarose percentage adequate for DNA size range may differ depending on agarose formulation. The total amount of agarose and 1× TAE buffer needed will vary based on the size of the gel and the electrophoresis equipment.
19. For higher percentage gels (>2%), it is better to heat the agarose at a lower heating power to prevent TAE evaporation. Higher percentage gels will also take longer to dissolve fully compared to lower percentage gels. It is recommended that mixtures are heated for an initial minute and then swirled for a few seconds before additional heating. The length of heating will vary based on the microwave used and should be optimized accordingly.
20. When handling higher percentage gels, it is best to add the appropriate amount of nucleic acid gel stain no longer than 5 min after the flask has been removed from the microwave. This is to ensure even incorporation of nucleic acid stain into the gel.
21. The period needed for the agarose gel to fully polymerize will vary depending on the gel's percentage. Higher percentage gels will polymerize faster and lower percentage gels will take longer to solidify. Ensure that gels are completely polymerized before removing the comb and gel cast. Precast gels may be stored overnight at 4 °C for running the next day. Wrap the gels in plastic film to avoid drying during storage. Make sure to let gels acclimatize to room temperature before use after retrieval from 4 °C storage.
22. If DNA loading dye was not added in PCR master mix, add DNA loading dye to samples before loading them into the wells.
23. Voltage and length of running may be adjusted based on equipment and band resolution needed. For example, separation of bands that are 20 base pairs apart in length will need to be ran the full gel length, whereas separation of bands a few hundred base pairs apart can be ran for half the length of a gel.

Acknowledgment

This work was supported by the Heart and Stroke Foundation of Canada (G-17-0018613), the Natural Sciences and Engineering Research Council of Canada (NSERC) (500865), the Canadian Institutes of Health Research (CIHR) (PJT-149046), and Operational Funds from the Hospital for Sick Children to P.D.-O.

References

1. Mouse Genome Sequencing Consortium, Waterston RH, Lindblad-Toh K, Birney E, Rogers J, Abril JF et al (2002) Initial sequencing and comparative analysis of the mouse genome. Nature 420:520–562
2. Saiki RK, Scharf S, Faloona F, Mullis KB, Horn GT, Erlich HA et al (1985) Enzymatic amplification of beta-globin genomic sequences and restriction site analysis for diagnosis of sickle cell anemia. Science 230:1350–1354
3. Saiki RK, Gelfand DH, Stoffel S, Scharf SJ, Higuchi R, Horn GT et al (1988) Primer-directed enzymatic amplification of DNA with a thermostable DNA polymerase. Science 239:487–491
4. Newton CR, Graham A, Heptinstall LE, Powell SJ, Summers C, Kalsheker N et al (1989) Analysis of any point mutation in DNA. The amplification refractory mutation system (ARMS). Nucleic Acids Res 17:2503–2516
5. Kim T (2000) PCR primer design: an inquiry-based introduction to bioinformatics on the World Wide Web. Biochem Mol Biol Edu 28:274–276
6. Ye J, Coulouris G, Zaretskaya I, Cutcutache I, Rozen S, Madden TL (2012) Primer-BLAST: a tool to design target-specific primers for polymerase chain reaction. BMC Bioinformatics 13:134
7. Pluthero FG (1993) Rapid purification of high-activity Taq DNA polymerase. Nucleic Acids Res 21:4850–4851
8. Qin J, Whyte WA, Anderssen E, Apostolou E, Chen HH, Akbarian S, Bronson RT, Hochedlinger K, Ramaswamy S, Young RA, Hock H (2012) The polycomb group protein L3mbtl2 assembles an atypical PRC1-family complex that is essential in pluripotent stem cells and early development. Cell Stem Cell 11:319–332

Check for updates

Chapter 2

Visualizing the Vascular Network in the Mouse Embryo and Yolk Sac

Anna R. Roy and Paul Delgado-Olguin

Abstract

Whole mount immunofluorescence is a valuable technique that can be used to visualize vascular networks in early developing embryonic tissues. This technique involves the permeabilization of fixed mouse embryos and yolk sacs, and primary antibody tagging of the endothelial cell marker platelet endothelial cell adhesion molecule 1 (Pecam-1). A secondary antibody tagged with a fluorophore targets the primary antibody, fluorescently labeling endothelial cells and revealing vascular networks.

Key words Whole mount immunofluorescence, Vascular network, Embryo, Yolk sac

1 Introduction

The vascular system develops early in life and is required for survival of the embryo [1–3]. Vascularization starts in the yolk sac, a membranous sac attached and surrounding the developing embryo, and later spreads to the embryo [1–3]. Early complications in vascularization can lead to developmental delays, stunted growth, abnormal vascular morphology, and in severe cases embryonic lethality [1–5]. Visualization of the vascular structure in the developing embryo is useful in the identification of vascular development abnormalities. One method for visualization of the developing vasculature in early mouse embryonic stages is whole mount immunofluorescence. Visualization of vascular structures in the whole mouse embryo and the yolk sacs is possible because the mouse embryo at developmental days (*E*) *E*7.5–10.5 is transparent [3, 4]. An advantage to this technique is that it is relatively easy to perform and does not require complex equipment. However, this technique uses fixed embryos and yolk sacs, and thus is limited to postmortem analysis.

Whole mount immunofluorescence is based on labeling a tissue-specific protein using an antibody to visualize a cell or tissue of interest [6]. Here we describe a method to visualize the vascular

Paul Delgado-Olguin (ed.), *Mouse Embryogenesis: Methods and Protocols*, Methods in Molecular Biology, vol. 1752, https://doi.org/10.1007/978-1-4939-7714-7_2,

system in the embryo and the yolk sac. Embryos or yolk sacs are fixed, permeabilized and incubated with a primary antibody targeting the platelet endothelial cell adhesion molecule 1 (Pecam-1), a marker of endothelial cells. Samples are then incubated with a secondary antibody linked to a fluorophore that targets the primary antibody. Imaging embryos or yolk sacs under fluorescent light allows the visualization of vascular networks.

2 Materials

Prepare all solutions in double distilled or ultrapure water.

2.1 Specimen Preparation

1. 1× Phosphate buffered saline (PBS): Prepare 10× PBS by dissolving 160 g of NaCl, 4 g of KCl, 28.8 g of Na_2HPO_4, and 4.8 g of KH_2PO_4 in 1.8 L of water. Adjust pH to 7.4 by adding 5 M NaOH solution. Add additional water up to 2 L and autoclave to sterilize. Store at room temperature. Prepare 1× PBS by adding 50 mL of 10× PBS in 450 mL of water. Autoclave to sterilize. Store at room temperature.
2. 90 mm Petri dish.
3. Tweezers.
4. 4% Paraformaldehyde (PFA): Prepare 20% PFA by adding 200 g of paraformaldehyde in 850 mL of water and 3 mL of 10 N NaOH. Heat solution in a water bath at 65 °C until clear. Cool down at room temperature and add water up to 1 L. Filter solution using sterile bottle-top filters. Aliquot into 50 mL conical tubes and store at −20 °C (*see* **Note 1**).
5. 1.5 mL clear microcentrifuge tubes.

2.2 Immunostaining

1. 1× Phosphate Buffered Saline Triton X-100 (PBST): 1× PBS with 0.2% Triton X-100. Store at 4 °C.
2. Permeabilization buffer: PBST with 0.1% bovine serum albumin (BSA) and 2% normal goat serum. Store at 4 °C (*see* **Note 2**).
3. Blocking buffer: 1× PBS with 0.1% BSA and 2% normal goat serum. Store at 4 °C.
4. Primary Antibody: mouse CD31/Pecam-1 (BD Pharmingen, 553370), dilute 1:100 in blocking buffer (*see* **Note 3**).
5. Secondary Antibody: Goat anti-mouse IgG (H + L) Highly Cross-Absorbed Secondary Antibody, Alexa Fluor 488 (Invitrogen Cat. No. A-11029), dilute 1:200 in blocking buffer.
6. 1.5 mL clear microcentrifuge tubes.
7. Tube rack.
8. Aluminum foil.
9. Laboratory orbital shaker.

2.3 Imaging

1. 1:1 glycerol solution: 50% glycerol in 50% PBS solution. Store at 4 °C (*see* **Note 4**).
2. 80% glycerol: glycerol diluted in PBS solution. Store at 4 °C.
3. 90 mm Glass petri dish with the bottom covered with black Sylgard (Dow Corning). Alternatively, dishes with the bottom covered with solidified agarose can be used. This will facilitate placement of the specimen for proper imaging.
4. Glass microscope slides and coverslips.

3 Methods

3.1 Sample Preparation

1. Animal procedures must follow Institutional guidelines and be approved by the corresponding Animal Care Committee. Euthanize pregnant mice between embryonic day (*E*) 7.5 and 10.5.
2. Cut the uterine horns to dissect the uterus with the embryos and place it into ice cold 1× PBS in a petri dish. Dissect out individual embryos from the uterus keeping the embryo, yolk sac, and placenta intact. Carefully remove the yolk sac by cutting the umbilical cord at the base of the placenta and place it into a 1.5 mL microcentrifuge tube in 1× PBS. Detach the embryo from the placenta by cutting the umbilical cord and place the embryos and yolk sacs into individual 1.5 mL tubes with 1× PBS. Please *see* **Note 5** if the embryo genotype is needed.
3. Remove the PBS from the tubes and fill them with 4% PFA and store overnight (O/N) at 4 °C (*see* **Notes 6** and **7**).

3.2 Immunostaining

1. Wash embryos/yolk sacs with 1 mL of 1× PBS 3× for 5 min at room temperature (RT).
2. Add 1 mL of permeabilization buffer and incubate at room temperature for 1 h.
3. Remove the permeabilization buffer.
4. Dilute the primary antibody (Pecam-1) in blocking buffer. Add 300 μL of primary antibody mix to each sample. Place tubes in a tube rack and on an orbital shaker set to speed 2 (low, approximately 30 rpm) and leave overnight at 4 °C.
5. Wash samples with 1× PBS 5× for 5 min at RT.
6. Add 500 μL of blocking buffer to the samples and incubate for 1 h at RT.
7. Dilute secondary antibody (Alexa-Fluor 488) in blocking buffer. Add 300 μL of secondary antibody mix to the samples. Cover the tubes with aluminum foil and place them on an orbital shaker set to speed 2 (approximately 30 rpm) for 1 h at

RT. Samples must be covered from light from this point on to avoid photobleaching.

8. Wash samples with 1 mL of 1× PBS 5× for 5 min each at RT.

3.2.1 Imaging: Embryos

1. After washing, remove the PBS and add 1 mL of 1:1 (50%) glycerol to the samples. Incubate for 3 h at 4 °C. Keep tubes covered with aluminum foil (*see* **Note 8**).
2. Remove 50% glycerol, add 1 mL of 80% glycerol and incubate for 1 h at 4 °C before imaging.
3. Transfer embryos into a 90 mm glass petri dish with the bottom covered with Black Sylgard, filled with 80% glycerol.
4. Optional: For proper positioning of embryos for imaging, cut a hole big enough to fit embryos in the Black Sylgard bottom using tweezers (*see* **Note 9**).
5. Image using a dissection microscope equipped with a fluorescent light lamp, or by confocal microscopy (*see* **Note 10**). The embryonic vascular network will be revealed in green fluorescence (Fig. 1).

3.2.2 Imaging: Yolk Sacs

1. After washing samples (**step 8** in Subheading 3.2), transfer the yolk sac with very little 1× PBS onto microscope slides.
2. Use tweezers to carefully flatten the yolk sac by making four half cuts to the center (Fig. 2).
3. Carefully place the coverslip on top and image immediately (*see* **Notes 10** and **11**). The yolk sac vascular network will be revealed in green fluorescence (Fig. 2).

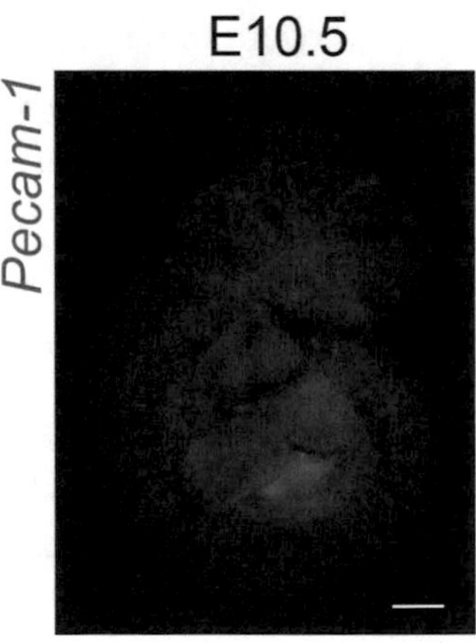

Fig. 1 Pecam-1-stained embryo at E10.5. The embryo was stained with a Pecam-1 primary antibody and a goat anti-mouse Alexa Fluor 488-coupled antibody. Scale bar = 100 μm

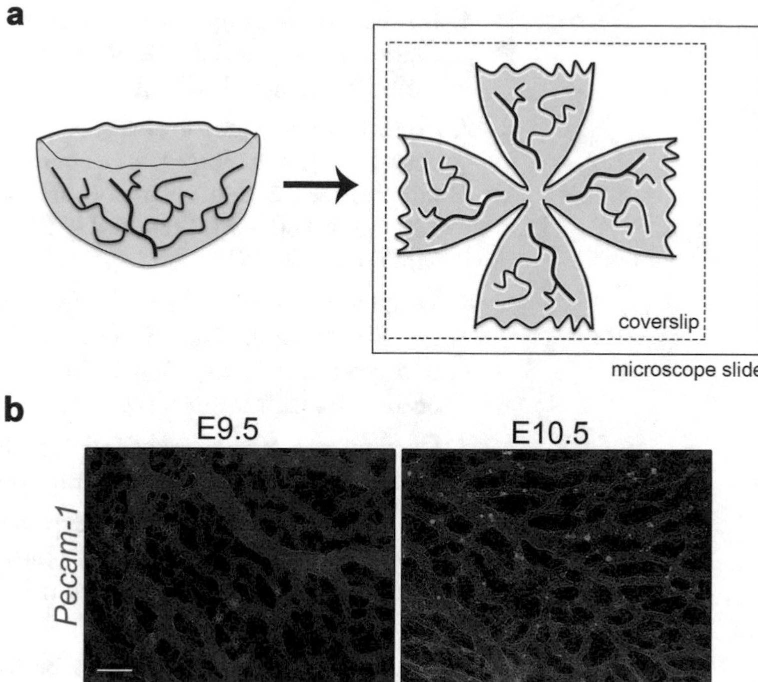

Fig. 2 Pecam-1 stained yolk sacs at *E*9.5 and *E*10.5. **(a)** Flattening the yolk sac for optimal visualization of the vasculature. Four cuts are made on the yolk sac using tweezers on a glass microscope slide. **(b)** Yolk sacs stained with a Pecam-1 primary antibody and a goat anti-mouse GFP secondary antibody. Scale bar = 100 μm

4 Notes

1. Paraformaldehyde is a very toxic and harmful irritant for eyes, skin and the respiratory tract. Wear proper protective equipment and work under a fume hood. Ensure that the PFA powder dissolves completely and allow the solution to cool down before filtration.
2. Ensure that the bovine serum albumin is inactivated before use. Inactivate by incubating in a water bath at 60 °C for 1 h. Aliquot into tubes and place in −20 °C.
3. It is best to perform a titration test with serial dilutions with untested antibodies.
4. Glycerol is a viscous substance. Vigorous shaking can cause bubbles to form. Ensure proper mixing of glycerol by gently flicking the tube side to side.
5. If genotyping of embryos is needed, a small piece of tissue (tail, limb buds, or tissues not to be analyzed by staining) can be dissected.

6. It is best to fix samples only for a few hours 4 h to O/N. Over fixation increases the risk of epitope masking and reduces the antibody–antigen binding [6].
7. Long term storage of embryos can be done at this stage by dehydrating the samples in a methanol (MeOH) series. Wash samples for 15 min each at room temperature in: 25%, 50%, 75%, and 100% MeOH in 1× PBS. Store samples in 100% MeOH at −20 °C.
8. Samples can be placed in 1:1 glycerol overnight or a week longer for clearing. The glycerol solution helps remove excess background staining. Depending on the staining, samples may need to be kept longer in glycerol solution.
9. Cutting an embryo -size hole in the Black Sylgard bottom facilitates positioning of the embryo for imaging.
10. Ensure that samples are not exposed to fluorescent light for long periods of time to prevent photo bleaching. For yolk sacs, slides can be covered by aluminum foil and stored for a week at 4 °C.
11. Pipetting a few μL of 1× PBS between the microscope slide and coverslip is important to prevent the yolk sac from being crushed by the coverslip.

Acknowledgment

This work was supported by the Heart and Stroke Foundation of Canada (G-17-0018613), the Natural Sciences and Engineering Research Council of Canada (NSERC) (500865), the Canadian Institutes of Health Research (CIHR) (PJT-149046), and Operational Funds from the Hospital for Sick Children to P.D.-O.

References

1. Marcel KL, Goldie LC, Hirschi KK (2013) Regulation of endothelial cell differentiation and specification. Circ Res 112(9):1272–1287
2. Park C, Kim TM, Malik AB (2013) Transcriptional regulation of endothelial cell and vascular development. Circ Res 112(10):1380–1400
3. Garcia MD, Larina IV (2014) Vascular development and hemodynamic force in the mouse yolk sac. Front Physiol 5:308
4. Coultas L, Chawengsaksophak K, Rossant J (2005) Endothelial cells and VEGF in vascular development. Nature 438(7070):937–945
5. Brindle NP, Saharinen P, Alitalo K (2006) Signaling and functions of angiopoietin −1 in vascular protection. Circ Res 98(8):1014–1023
6. Ursini-Siegel J, Beauchemin N (eds) (2016) The tumor microenvironment: methods and protocols. Meth Mol Biol 1458. https://doi.org/10.1007/978-1-4939-3801-8_1

Chapter 3

In Vivo Evaluation of the Cardiovascular System of Mouse Embryo and Fetus Using High Frequency Ultrasound

Yu-Qing Zhou, Lindsay S. Cahill, and John G. Sled

Abstract

Genetically engineered mice have been widely used for studying cardiovascular development, physiology and diseases. In the past decade, high frequency ultrasound imaging technology has been significantly advanced and applied to observe the cardiovascular structure, function, and blood flow dynamics with high spatial and temporal resolution in mice. This noninvasive imaging approach has made possible longitudinal studies of the mouse embryo/fetus in utero. In this chapter, we describe detailed methods for: (1) the assessment of the structure, function, and flow dynamics of the developing heart of the mouse embryo during middle gestation (*E*10.5–*E*13.5); and (2) the measurement of flow distribution throughout the circulatory system of the mouse fetus at late gestation (*E*17.5). With the described protocols, we are able to illustrate the main cardiovascular structures and the corresponding functional and flow dynamic events at each stage of development, and generate baseline physiological information about the normal mouse embryo/fetus. These data will serve as the reference material for the identification of cardiovascular abnormalities in numerous mouse models with targeted genetic manipulations.

Key words Mouse, Embryo, Fetus, Ultrasound, Doppler, Heart, Vessel, Flow, Function

1 Introduction

With the rapid progression of genetic technologies, mice have become the major animal models of cardiovascular development and physiology, providing a platform for exploring the role of genes in normal development, malformations, and diseases. Many studies have been conducted on cardiac development using mouse embryos [1, 2]. From day 8.5 post-implantation (*E*8.5), when a primitive heart tube has formed and starts to pump blood, to day 13.5 of gestation (*E*13.5), when heart formation is mostly complete, the mouse embryonic heart undergoes a series of dramatic structural changes similar to those in humans. Defects in major developmental events such as heart tube looping, chamber formation and septation, trabeculation, valvular formation, and outflow tract division at consecutive stages, are associated with characteristic forms of congenital heart disease. The relatively short period of

Paul Delgado-Olguin (ed.), *Mouse Embryogenesis: Methods and Protocols*, Methods in Molecular Biology, vol. 1752, https://doi.org/10.1007/978-1-4939-7714-7_3, © Springer Science+Business Media, LLC, part of Springer Nature 2018

time for heart development in the mouse is convenient for studying the genotype–phenotype relationships that determine normal heart development and cardiac malformation following genetic mutation [3]. However, much of the research on structural changes that occur as the heart develops uses ex vivo histological examinations. In vivo observation of the cardiac function and blood flow dynamics corresponding to the anatomical changes at various time points throughout development can significantly enhance our understanding of cardiac development and malformation. Many aspects of heart development such as the effect of the blood flow shear stress on gene expression of developing structures [4], valvular morphology and local flow pattern, interaction between atrium and ventricle, developmental changes of ventricular systolic and diastolic function, as well as the pathophysiology of cardiac malformations will be more comprehensively elucidated by in vivo observation. Furthermore, as cardiogenesis is critical for embryo survival, and defective cardiovascular development is one of the main causes of in utero lethality in mutant mouse embryos/fetuses, it is important to use in vivo observation to accurately determine the developmental stage at which defects arise and when lethality occurs. The developmental timing of lethality is usually a good indicator of the type of cardiovascular defect present [5]. By using a noninvasive technology, the longitudinal monitoring of the entire litter and/or individual embryos/fetuses of special interest provides valuable information about the dynamic change of litter size, and might also suggest underlying causes.

The mouse fetus is useful for studying the physiology of the entire circulatory system. For instance, the mouse fetus demonstrates "brain sparing" [6], a physiological response to hypoxic stress that is also seen in large animal and human fetuses. When the blood flow or oxygen supply to the fetus is reduced, the fetal circulatory system has the capability to redistribute flow to maintain the oxygen supply to vital organs including the heart and brain at the expense of other organs such as lung and liver [7]. The flow redistribution throughout the fetal vascular system occurs mainly by autonomic regulation of flow across three shunts: ductus venosus, foramen ovale, and ductus arteriosus [8–10]. Mutant mice with placental dysfunction have also been used to study the flow dynamics in placental–fetal circulation and mechanisms related to intrauterine growth restriction [11, 12]. Furthermore, mouse fetuses with gene mutations have been evaluated under challenge conditions such as hypoxia and hyperoxia [13, 14], or exposure to tobacco smoke, caffeine, and alcohol, to elucidate gene functions in related pathophysiological processes and disease development [15–17]. The growing use of mutant mouse models generates a great need for noninvasive in vivo assessment of mouse fetal circulatory physiology.

High-frequency ultrasound technology was introduced for imaging mouse embryos in the mid-1990s and since then has been frequently used to assess the cardiovascular system of the mouse embryo/fetus [18–22]. As a noninvasive technology, it has offered a reproducible and a readily available approach for in vivo longitudinal observation of cardiovascular morphology, function, and flow dynamics, with little perturbation to the embryonic/fetal physiology. Technological advances in this technology during the past decade include high frequency linear array transducers and Doppler color flow mapping [23]. Linear array transducers provide better imaging quality and much higher temporal resolution. Doppler color flow mapping is especially useful in differentiating embryonic/fetal cardiac structures and small vessels by sensitively visualizing the blood flow in the cardiac chamber and vascular lumen, and also in guiding quantitative measurement of flow velocity using pulsed Doppler [24]. Benefitting from these technological advancements, our ability to image the mouse embryonic/fetal cardiovascular system has been significantly improved, therefore related methods need to be updated.

In this chapter we describe detailed methods for: (1) the assessment of the structure, function and flow dynamics of the developing heart of the mouse embryo during mid-gestation (*E*10.5–*E*13.5); and (2) the measurement of flow distribution throughout the circulatory system of the mouse fetus at late gestation (*E*17.5). With the described protocols, we are able to obtain baseline information for normal cardiovascular development and the physiology of the mouse embryo/fetus in utero, and to provide reference data for future studies on cardiovascular abnormalities in numerous mouse models with targeted genetic modifications.

2 Materials

1. Animals: Mice of selected wild-type or mutant strains at *E*10.5–*E*17.5 are suitable for the described protocols. During the embryonic period from *E*10.5–*E*13.5 the dramatic structural changes and corresponding flow dynamics of the developing heart are to be observed. During the fetal period from *E*13.5 to term (*E*18.5 for most mouse strains), the flow distribution throughout the fully developed circulatory system can be evaluated. This is illustrated in this chapter using the *E*17.5 fetus. The described procedures are performed on live animals, thus the experimental protocol should be approved by the institutional Animal Care and Use Committee.
2. High-frequency ultrasound imaging system: A high frequency ultrasound system that combines anatomical and blood flow imaging capabilities is needed. Here, as an example of such a

system, we describe the characteristics and modes of operation of the Vevo2100 (VisualSonics Inc., Toronto, Canada) with a 40 MHz linear array transducer. Four standard modes are used for cardiovascular evaluation: (a) B-mode: The two-dimensional (2D) greyscale anatomical imaging has a spatial resolution of about 100 μm laterally and 50 μm axially. The maximum field-of-view is 14 mm wide and 15 mm deep, with the image width covering about half of the maternal abdomen of a pregnant mouse and sufficient penetration depth to image the deepest embryo/fetus with acceptable signal intensity. With the current linear array transducer, the real-time imaging frame rate can be close to 1000 fps when image width is reduced to the minimum. (b) M-mode: M-mode imaging is used for recording dynamic changes in position and dimension of the heart or vessel of interest along a selected direction over a cardiac cycle enabling diameter and functional parameters to be measured and derived. (c) Doppler color flow mapping: Doppler color flow mapping visualizes the blood flow and its direction using pseudocolors (red and blue) superimposed on the B-mode image. Color flow mapping qualitatively demonstrates the velocity distribution across the heart chamber, orifice or vascular lumen, identifies the sites with relatively high flow velocity or turbulence, and guides positioning of pulsed wave Doppler sample volume for velocity measurement. This modality is especially important in mouse embryonic/fetal imaging because of the small size, close proximity, and weak echogenicity of the vasculature in B-mode tissue imaging. (d) Pulse wave Doppler: With the guidance of 2D tissue imaging or Doppler color flow mapping, pulsed Doppler is used for recording the blood flow velocity spectrum at a site of interest. Doppler sample volume size is adjustable according to the dimension of the targeted vessels. Doppler measurement needs to be angle-corrected to give accurate velocity estimate and it is advisable to keep the angle between the ultrasound beam and flow direction less than 60°. The recorded flow spectrum is used either for qualitative assessment of flow status or for quantitative calculation of volume flow in cases where the cross-sectional area of the vessel is also measured. A comprehensive evaluation of the embryonic/fetal cardiovascular system typically needs all four of the functional modalities described above.

3. Imaging station: As an important accessory to the ultrasound system, the imaging station includes a platform with temperature adjustment, heart rate and body temperature monitoring, and a stand for holding the transducer (VisualSonics Inc., Toronto, Canada). The embryo/fetus has a random orientation inside the maternal abdomen, but a strict spatial relationship is needed for visualizing a certain structure and measuring diame-

ter and flow velocity. That situation demands that both the stand holding the transducer and the imaging platform supporting the mouse have sufficient freedom for changing their orientations or angles. When the spatial relationship between transducer and targeted embryonic/fetal structure is decided, the positions of the transducer stand and the imaging platform need to be locked to ensure a stable recording. With microimaging of the mouse embryo/fetus, it is impossible to get proper data by hand holding the ultrasound transducer because a very small hand movement will cause significant changes in the image. Fine adjustment of the imaging platform or transducer holder is often needed to further optimize image details or get back to the original structure after the embryo/fetus has moved.

4. Anesthetics: isoflurane.
5. Hair-removing cream: It is used to remove the hair from the abdomen to be imaged.
6. Warmed ultrasound gel: It is for coupling the ultrasound transducer with the mouse body.

3 Methods

3.1 Pre-imaging Preparation

1. Place a pregnant mouse into an anesthesia induction chamber and set the level of isoflurane at 5% in medical air (21% oxygen). It takes about 3–5 min for the mouse to become unconscious and immobile.
2. Move the anesthetized mouse onto an imaging platform with the temperature set at 40–42°. A thermal probe is inserted into the mouse rectum for measuring the maternal body temperature (maintained at ~37° throughout the imaging session). If necessary, an extra heating lamp can be used. The anesthesia is maintained with isoflurane at 1.5% in 21% oxygen by face mask.
3. Tape the mouse paws on top of the four ECG leads for monitoring maternal heart rate.
4. Apply hair removing cream to the whole abdomen of the pregnant mouse to cleanly remove hair.
5. Put sufficient ultrasound gel on the mouse abdomen and avoid any air bubbles inside the gel.
6. Orient the ultrasound transducer for a transverse imaging section and scan over the whole abdomen of the pregnant mouse, from the lower to upper parts and from one side to another, to estimate the approximate number of embryos/fetuses.
7. Locate the embryo/fetus of interest, determine the orientation of the targeted embryo/fetus inside the maternal abdomen and identify the corresponding appearance in the image on the screen of the ultrasound system (*see* **Notes 1–3**).

3.2 Imaging the Developing Heart of the Mouse Embryo During Middle Gestation

1. During the period from *E*8.5 to *E*13.5, dramatic anatomical and functional changes of the developing heart can be observed using ultrasound imaging. In our previous study using a 55 MHz single element transducer (Vevo660, 770, VisualSonics Inc. Toronto) on an exteriorized uterine horn, a flow velocity waveform was detected using pulsed Doppler at *E*8.5, the first day the primitive heart tube started to pump blood through the embryo. At *E*9.5 a U-shaped tubular primitive heart was observed. The Doppler blood velocity waveform in the atrio-ventricular canal had a double peak with velocity up to 3 cm/s. The Doppler waveform in the outflow tract had a large antegrade flow during systole and small retrograde flow during diastole [19]. In our current studies using a 40 MHz linear array transducer with advanced Doppler color flow mapping (Vevo2100), embryonic cardiac structure, function and blood flow can be readily evaluated from *E*10.5 onwards. Exteriorization of the uterine horn is not necessary and therefore the perturbation to the embryonic physiology is avoided. Under high frequency ultrasound, mouse embryonic blood looks bright compared to the surrounding tissues, especially during mid gestation, providing extra advantages in several aspects (*see* **Note 4**).
2. At *E*10.5, the ultrasound transducer is oriented to get a sagittal section of the embryo (Fig. 1a). In this view, a folded heart tube can be delineated, with the inflow tract on the caudal side, the outflow tract on the cranial side, and the ventricular segment bulging anteriorly to the ventral side of the embryo. Real-time B-mode imaging shows the hyperechogenic blood cells moving toward the ventricular segment through the inflow tract during diastole and out from the ventricular segment through the outflow tract during systole. Doppler color flow mapping more clearly indicates the flow directions with different colors (Fig. 1b, d). Because the blood flow velocity in the embryo is very low, several settings (including velocity scale, gain and wall filter levels) need to be adjusted to optimize the Doppler color flow mapping (*see* **Note 5**). Pulsed Doppler from the inflow tract demonstrates a double-peaked velocity waveform, with a lower early diastolic wave (*E* wave) probably caused by the active relaxation of the ventricular segment, and a relatively higher late diastolic wave (*A* wave) pos-

Fig. 1 (continued) and nucleated red blood cells and high resolution of high frequency ultrasound. (**b**) Doppler color flow mapping showing the diastolic inflow (red) through inflow tract toward the ventricle. (**c**) Pulsed Doppler waveforms at inflow tract, showing an early diastolic waveform (*E* wave) caused by ventricular relaxation and a late diastolic waveform (*A* wave) caused by atrial contraction. Retrograde flow during systole is detected as the negative waveforms under baseline. (**d**) Doppler color flow mapping showing the systolic flow (blue) through outflow tract away from the ventricle. (**e**) Pulsed Doppler waveforms at the outflow tract, showing a systolic waveform. Retrograde flow during diastole is detected as the positive waveforms above baseline. *Br* brain

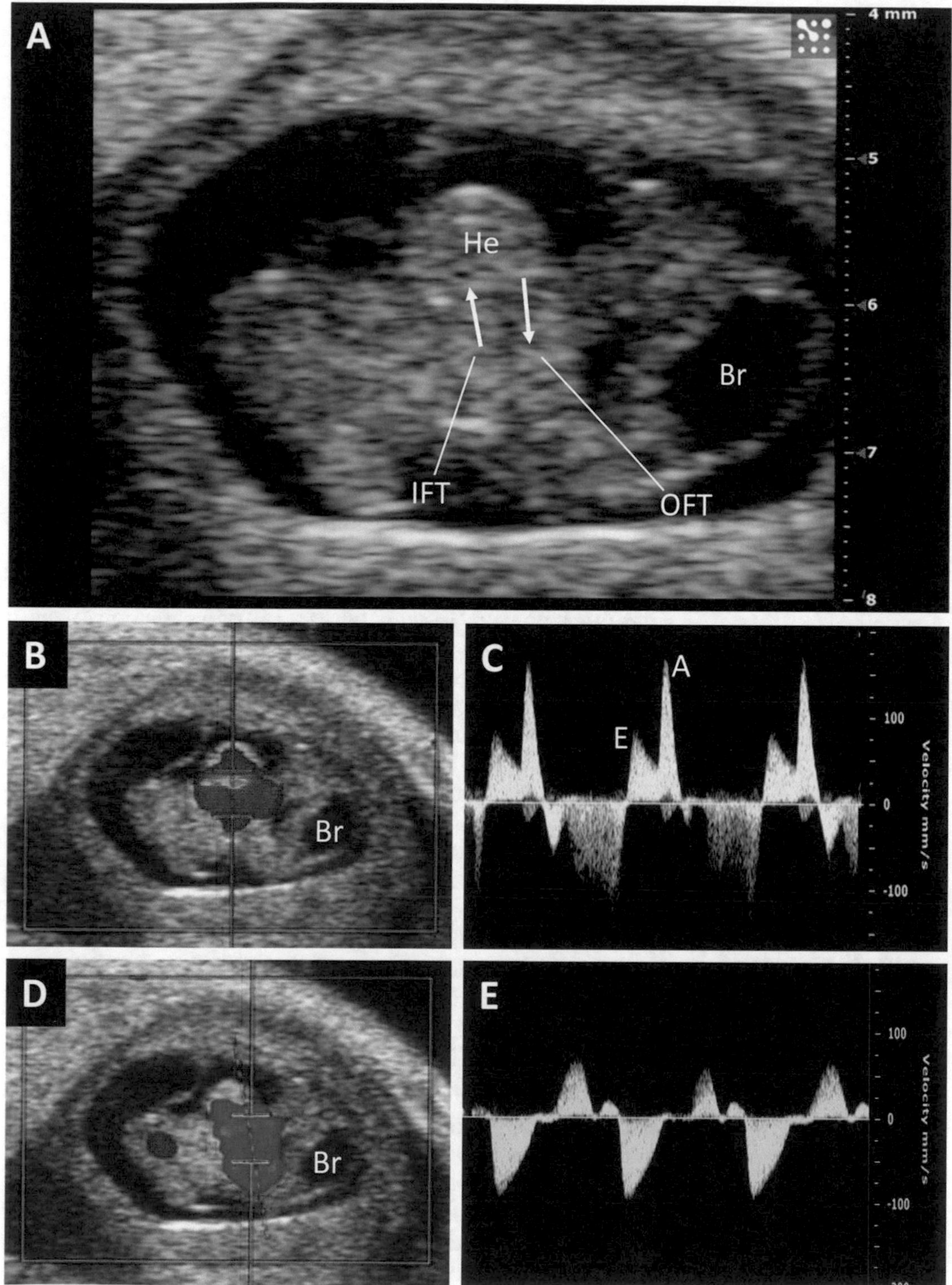

Fig. 1 Representative images of embryo at *E*10.5. (**a**) A sagittal section of embryo, with the primitive heart tube looping and bulging anteriorly to the ventral side. The inflow tract (IFT) to and outflow tract (OFT) away from the heart (He) are clearly visualized. In real-time imaging the movement of blood flow was visible because of the large

sibly caused by the contraction of the atrial segment. The negative waveform during systole is assumed to be due to the lack of valvular structure to prevent retrograde flow at this gestational age (Fig. 1c). Pulsed Doppler from the outflow tract presents a single-peaked velocity waveform, with blood flow moving away from the heart during systole. The retrograde waveform during diastole may be due to the lack of valvular structure to prevent reverse flow back to the ventricular segment at this stage (Fig. 1e).

3. At *E*11.5, the embryonic heart starts to show a shape of four chambers but without septation. A transverse imaging section of the embryo is more favorable to visualize the common atrium, common ventricle, common atrioventricular canal, and common outflow tract (Fig. 2a, b). Doppler color flow mapping clearly indicates the inflow through the common atrioventricular canal (Fig. 2c) and the outflow through the common outflow tract (Fig. 2e). Pulsed Doppler demonstrates typical inflow (Fig. 2d) and outflow (Fig. 2f) waveforms similar to those seen at *E*10.5.

4. At *E*12.5, the interventricular septum is forming but septation has not completed yet. A transverse imaging section is more favorable for visualizing the four chambers of the heart. On real-time B-mode imaging, the septum is the dark region between the left and right ventricles relative to the brighter blood cells in heart chambers. During diastole the flow streams in the left and right ventricular inflow tracts (future mitral and tricuspid orifices) are able to be partially differentiated. During systole the flow streams from the left and right ventricles lead into the common outflow tract which has not been completely separated into the aorta and pulmonary artery yet. Doppler color flow mapping clearly visualizes the diastolic inflows in both the left and right ventricular inflow tracts, but without complete separation, and the systolic outflow merged from both left and right ventricles (Fig. 3a–c). Pulsed Doppler shows symmetrical diastolic inflow waveforms between the two ventricles, with low *E* wave and high *A* wave (Fig. 3d–g), and a dominantly antegrade waveform from the outflow tract (Fig. 3h, i).

5. At *E*13.5, embryonic cardiac development has completed, showing complete ventricular septation and valvular leaflets. On B-mode imaging in a transverse section of the embryo, the four cardiac chambers with complete interventricular septum, mitral and tricuspid valves, and flow streams in two separated ventricular inflow tracts are observed. Slightly above the level of the four chamber section, the separated pulmonary artery and aorta can be visualized (Fig. 4a, b). By manipulating the orientation of the transducer and making the ultrasound beam direction perpendicular to the interventricular septum,

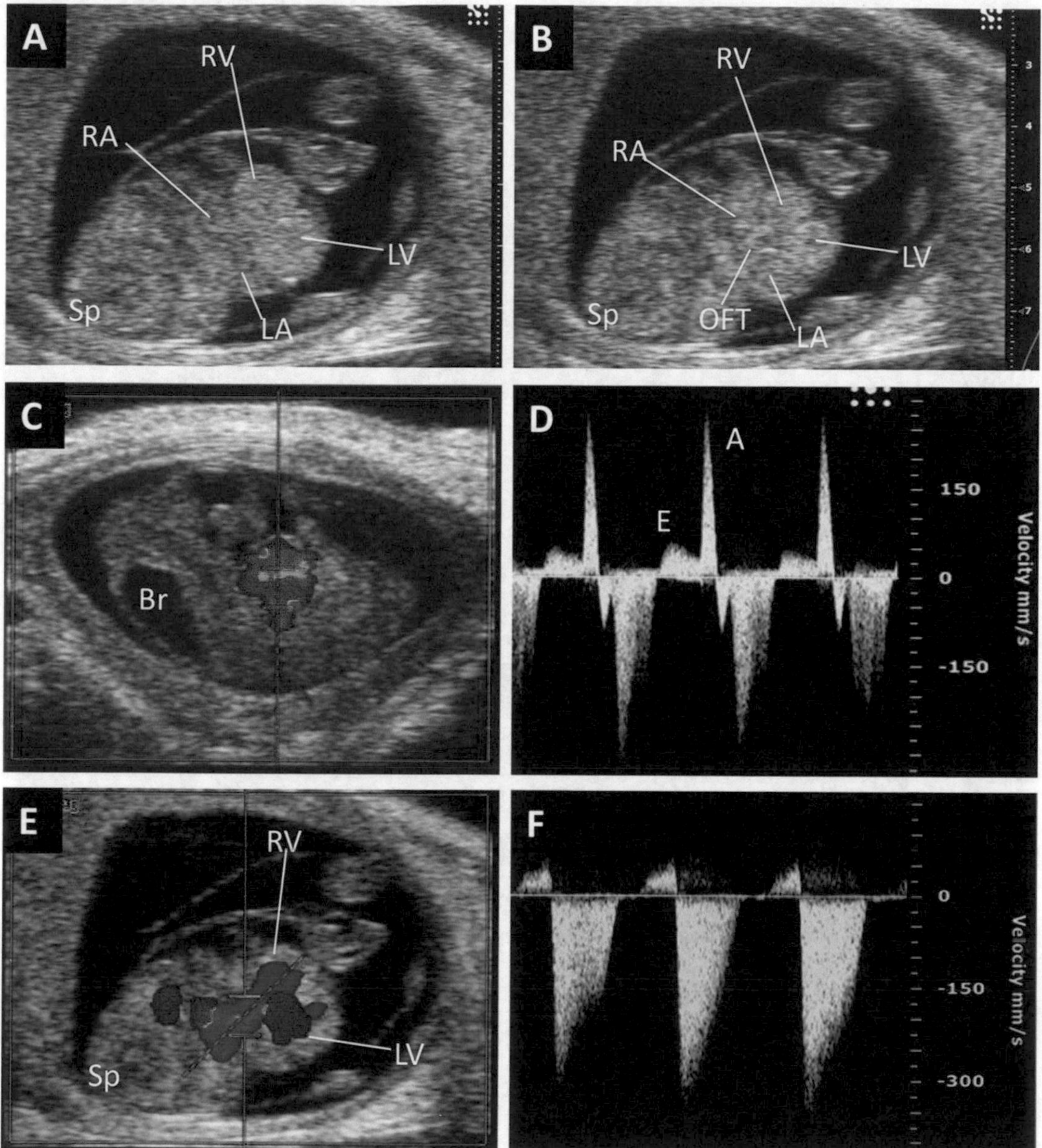

Fig. 2 Representative images of embryo at *E*11.5. (**a**) and (**b**) are transverse sections of the embryo, with four cardiac chambers delineated at diastole and systole, respectively. Four cardiac chambers are not separated as showed in panel **a**, and two ventricles have a common outflow tract (OFT) as seen in panel **b**. (**c**) Doppler color flow mapping showing the diastolic inflow (red) through common inflow tract toward the ventricle in a sagittal section of embryo. (**d**) Pulsed Doppler waveforms at inflow tract, showing an early diastolic *E* wave and a late diastolic *A* wave. (**e**) Doppler color flow mapping showing the flow (blue) through the common outflow tract away from heart during systole. (**f**) Pulsed Doppler waveforms of the systolic outflow, demonstrated as the negative waveforms under baseline. *Br* brain, *LA* left atrium, *LV* left ventricle, *RA* right atrium, *RV* right ventricle, *Sp* spine

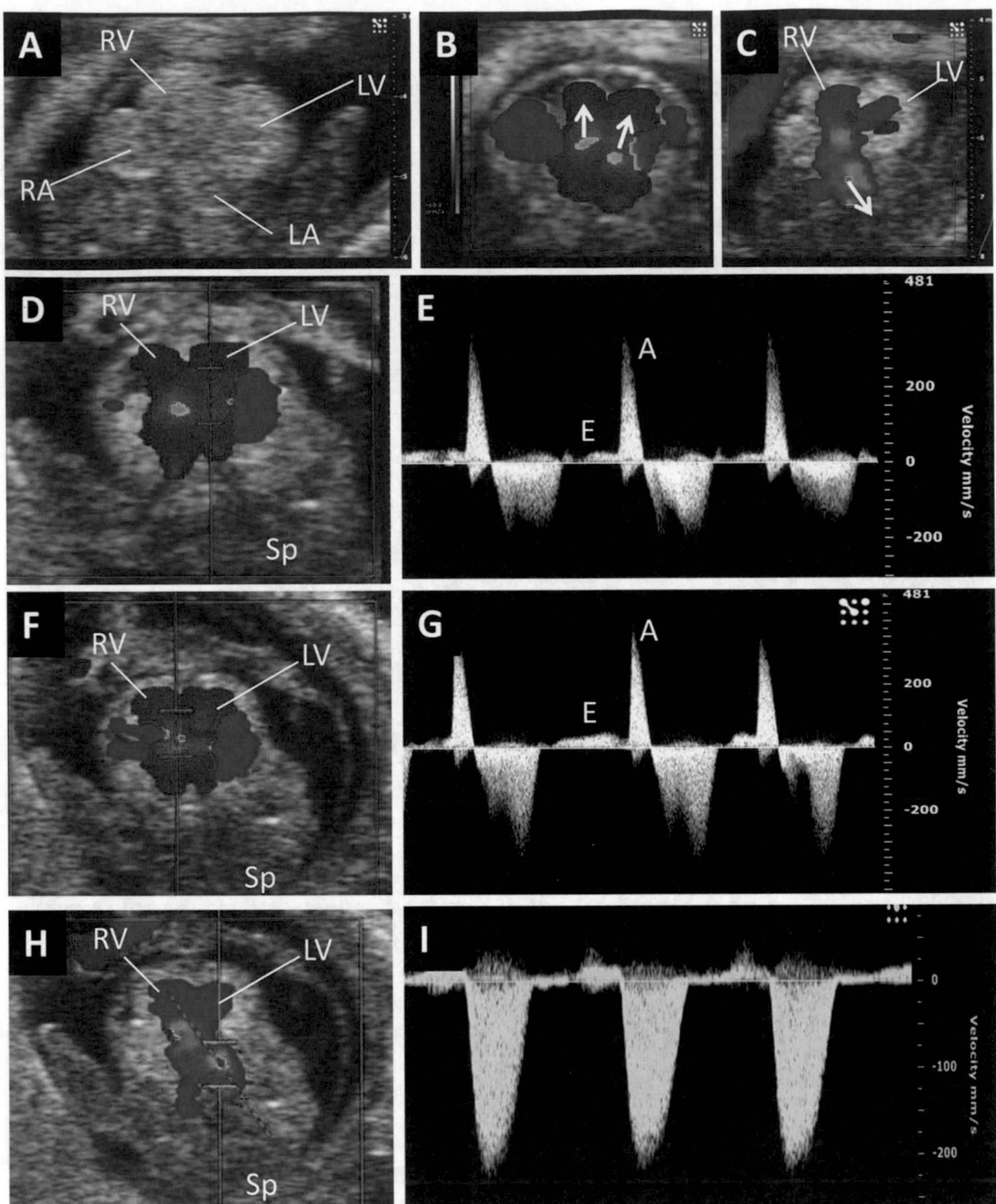

Fig. 3 Representative images of embryo at *E*12.5. (**a**), (**b**), and (**c**) are transverse sections of the embryo with four cardiac chambers, inflow (red) at diastole and outflow (blue) at systole, respectively. At *E*12.5, four cardiac chambers still communicate with each other, and the outflow tract is not completely divided. (**d**) Doppler color flow mapping showing the diastolic inflow (red) with pulsed Doppler sample volume at left ventricular inflow tract. (**e**) Pulsed Doppler waveform of left ventricular inflow showing a small *E* wave and a large *A* wave. (**f**) Doppler color flow mapping showing the diastolic inflow (red) with pulsed Doppler sample volume at right ventricular inflow tract. (**g**) Pulsed Doppler waveform of right ventricular inflow showing a small *E* wave and a large *A* wave. (**h**) Doppler color flow mapping showing the systolic outflow (blue) with pulsed Doppler sample volume at ventricular outflow tract (close to future main pulmonary artery). (**i**) Pulsed Doppler waveform of ventricular outflow showing a single waveform away from the heart. See previous figures for other abbreviations

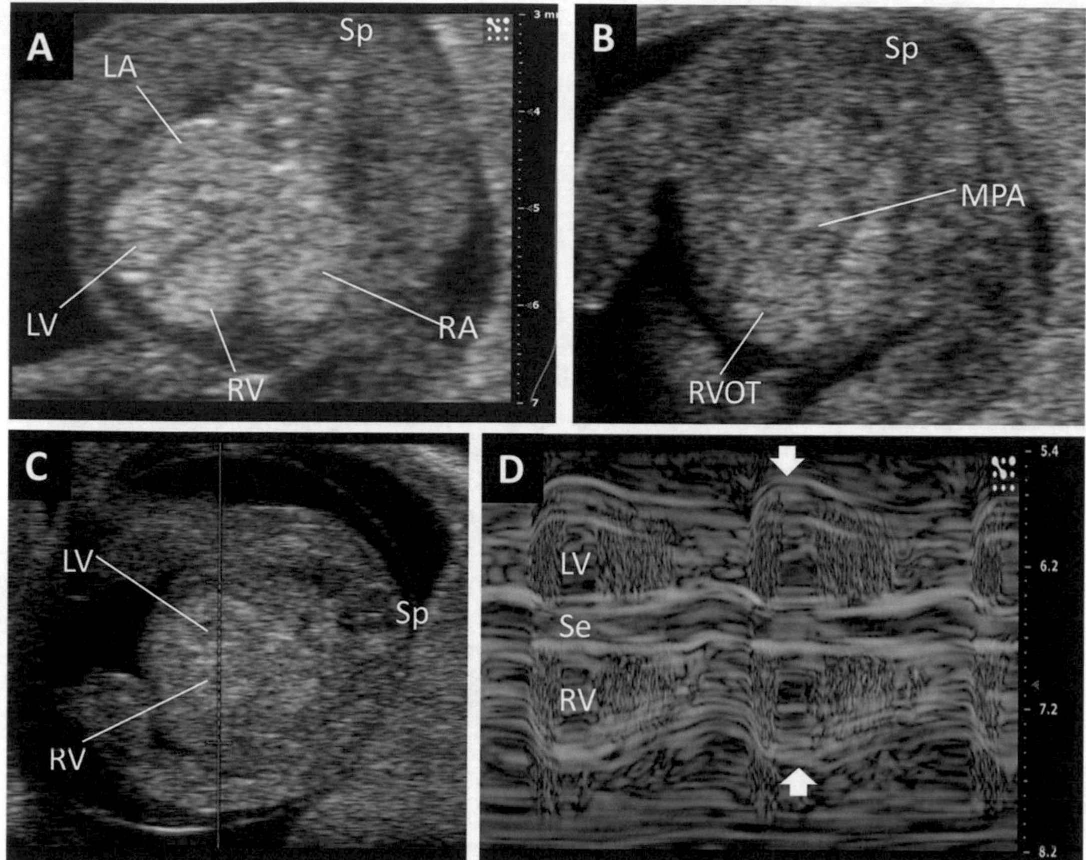

Fig. 4 Representative images of embryo at *E*13.5. (**a**) A transverse section of the embryo showing the four cardiac chambers, with left and right ventricles (LV, RV) completely separated by ventricular septum. Left and right atria (LA, RA) communicate with each other, as manifested by the movement of blood cells through the atrial septum in real-time B-mode imaging. (**b**) A transverse section of the embryo (at the level of cardiac base) showing right ventricular outflow tract (RVOT) leading to the main pulmonary artery (MPA). (**c**) B-mode image for guiding the M-mode recording of left and right ventricles, with a perpendicular interception angle between ultrasound beam and ventricular septum. (**d**) M-mode recording of left and right ventricles and interventricular septum (Se). Arrows indicate the borders (epicardium) of left and right ventricles with the maximal dimensions during diastole. See previous figures for other abbreviations

M-mode is recorded to demonstrate the dynamic changes of the left and right ventricular dimensions and septal thickness over a cardiac cycle (Fig. 4c, d) for estimating ventricular systolic function (*see* **Note 6**).

6. In a four chamber section, Doppler color flow mapping shows two separated and parallel inflow streams through the mitral and tricuspid orifices. Pulsed Doppler detects symmetrical diastolic inflow waveforms between the two ventricles (Fig. 5a–d), which will be used for assessing the left and right ventricular diastolic function (*see* **Note 7**) [25].

7. In the transverse section at the level of the cardiac base, Doppler color flow mapping shows that the aorta and pulmonary artery cross each other, and pulsed Doppler demonstrates similar outflow waveforms (Fig. 5e–h). The Doppler flow velocity recordings from the aorta and pulmonary artery are used for calculating cardiac output. More details will be given in the following section.

3.3 Evaluation of the Fetal Heart and the Flow Distribution Throughout the Fetal Circulatory System During Late Gestation

1. Fetal cardiac structures: Following the completion of cardiac formation on *E*13.5, comprehensive evaluation of cardiac structure, function, and flow dynamics can be conducted. The four chamber view of the fetal heart is always one of the most favorable imaging sections. With the fetal orientation in the operator's mind, the ultrasound transducer and imaging platform are manipulated to get a transverse imaging section through the fetal chest. With about 45° to the longitudinal axis of the fetus, a four-chamber view of the fetal heart can be obtained. Further fine adjustments are usually needed for optimizing the images for specific structures. As the apex of the fetal heart points to the left side, the left and right ventricles can be identified accordingly. With B-mode imaging, the left and right atria, the left and right ventricles, the mitral and tricuspid valves, and the continuity of interatrial and interventricular septa are assessed. Slightly moving the imaging section to the level of the cardiac base with further optimization of the imaging angle (aligning the imaging section with the longitudinal axis of the targeted vessel), the aortic and pulmonary orifices and the blood flow streams through them are evaluated. Doppler color flow mapping is essential for visualizing the normal blood flow streams through the mitral, tricuspid, aortic, and pulmonary orifices, and for identifying individual vessels especially when the targeted vessels are not very clear in the B-mode tissue image due to the low echogenicity of fetal vasculature and the close proximity of several vessels (*see* **Note 5**). Doppler color flow mapping also helps for angle correction of flow velocity measurements by pulsed Doppler. In cases with cardiac abnormalities such as valvular stenosis, regurgitation,

Fig. 5 (continued) showing the diastolic inflow (blue) with pulsed Doppler sample volume at right ventricular inflow tract. (**d**) Pulsed Doppler waveforms of right ventricular inflow showing a small *E* wave and a large *A* wave. (**e**) Doppler color flow mapping showing the systolic outflow (red) with pulsed Doppler sample volume at proximal aorta. (**f**) Pulsed Doppler waveforms of aortic flow. (**g**) Doppler color flow mapping showing the systolic outflow (red) with pulsed Doppler sample volume at proximal main pulmonary artery (MPA). (**h**) Pulsed Doppler waveforms of main pulmonary arterial flow. Note that the aorta and main pulmonary artery cross each other in the Doppler color flow mapping in the transverse imaging section at the level of the cardiac base, with the aorta starting from left ventricle and moving toward the right side of the midline, and the main pulmonary artery originating from the right ventricle and pointing to the descending thoracic aorta on the left side of spine (Sp) as illustrated in panels (**e**) and (**g**). See previous figures for other abbreviations

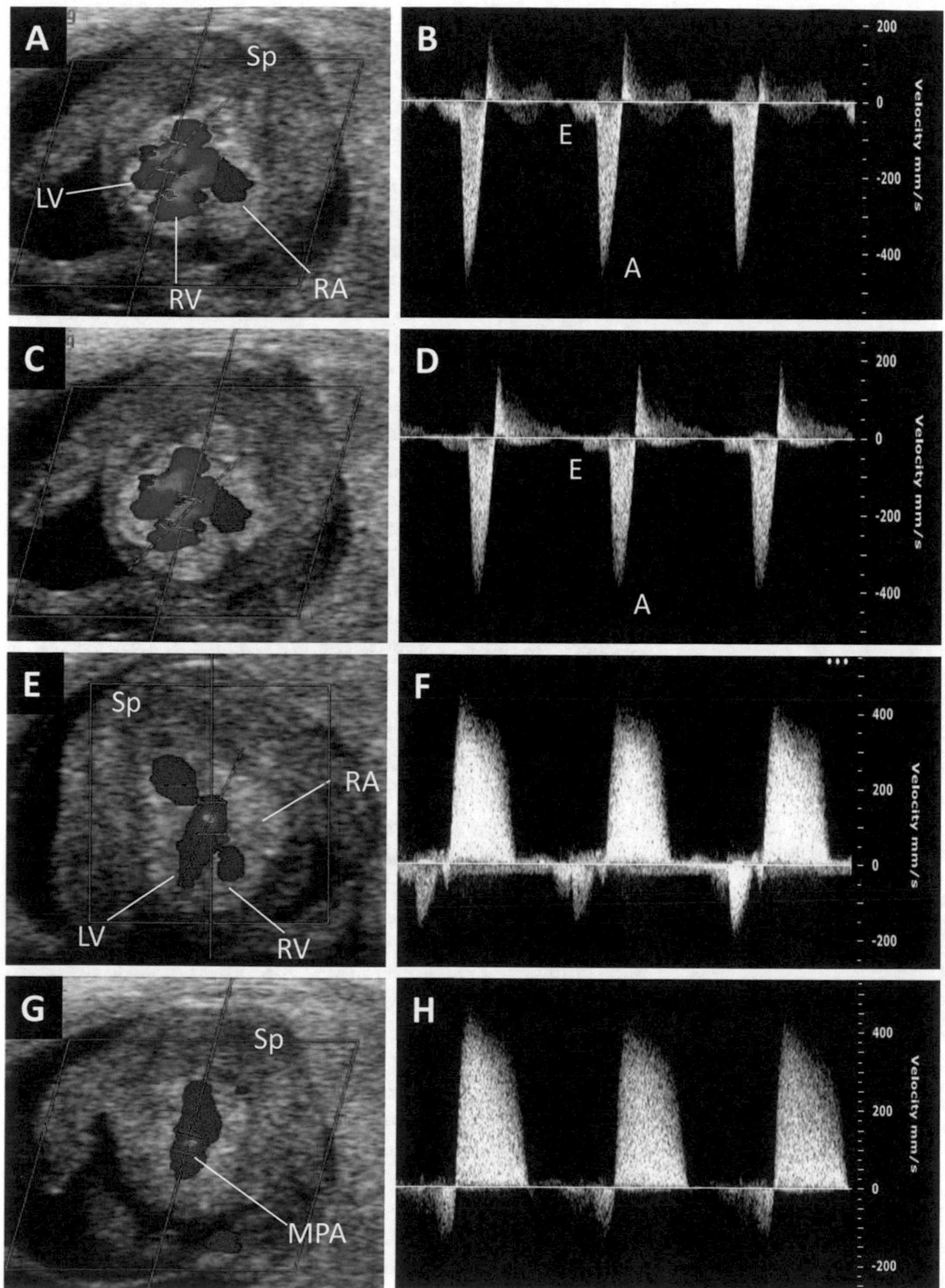

Fig. 5 Representative Doppler flow recordings of embryo at *E*13.5. (**a**) Doppler color flow mapping showing the diastolic inflow (blue) with pulsed Doppler sample volume at left ventricular inflow tract. (**b**) Pulsed Doppler waveforms of left ventricular inflow showing a small *E* wave and a large *A* wave. (**c**) Doppler color flow mapping

vessel narrowing, septal defects, and flow shunts, Doppler color flow mapping is especially useful for locating the structural abnormality because of significantly higher flow velocity caused by defective cardiac structures. By adjusting the pulse repetition frequency (velocity scale) the location of the highest flow velocity can be easily visualized in color flow mapping because of Doppler aliasing (*see* **Note 5**). Pulsed Doppler measurement is used to further confirm the abnormal flow velocity [26]. However, this chapter does not discuss in detail the identification of cardiac structural abnormalities and the consequent flow dynamics in mutant mice.

2. Fetal cardiac function: The ventricular dimensions and systolic function by M-mode, and the ventricular diastolic function by pulsed Doppler can be followed up through the rest of gestation. The details have been discussed above.

3. Foramen ovale: The foramen ovale is a normal shunt in the fetal heart, allowing the highly oxygenated blood flow returning from the ductus venosus to pass into the left atrium. In a four-chamber section of the heart, Doppler color flow mapping is used to visualize the shunting flow stream through the foramen ovale, and pulsed Doppler flow velocity spectrum is recorded to the left side of the atrial septum. However, it is difficult to measure the dimension of the foramen ovale for direct assessment of the shunting flow magnitude. The flow through the foramen ovale is calculated from the flow measurements of the ascending aorta and pulmonary blood flow (described below) [24].

4. Ascending aorta: From the four chamber view, the imaging section is slightly moved toward the basal level of the fetal heart and tilted anteriorly (toward the coronal plane of the fetus) to visualize the left ventricular outflow tract, aortic orifice and ascending aorta. With Doppler color flow imaging, the blood flow from the left ventricle through the aortic orifice to the ascending aorta is visualized, and the highest velocity can be identified at the level of the aortic valves as the brighter color or Doppler aliasing (Fig. 6a). For flow velocity measurement, the intercept angle between the longitudinal axis of the color flow and the ultrasound beam should be as small as possible (never larger than 60°). The pulsed Doppler sample volume is placed at the ascending aorta distal to the aortic sinus for recording the velocity waveforms. The Doppler sample volume should be large enough to cover the whole aortic lumen. To measure the aortic diameter for calculating the aortic flow, that is the left ventricular output, the transducer orientation is adjusted to achieve a perpendicular interception between the ascending aorta and the ultrasound beam within the existing imaging section. An M-mode cursor line is placed at the level

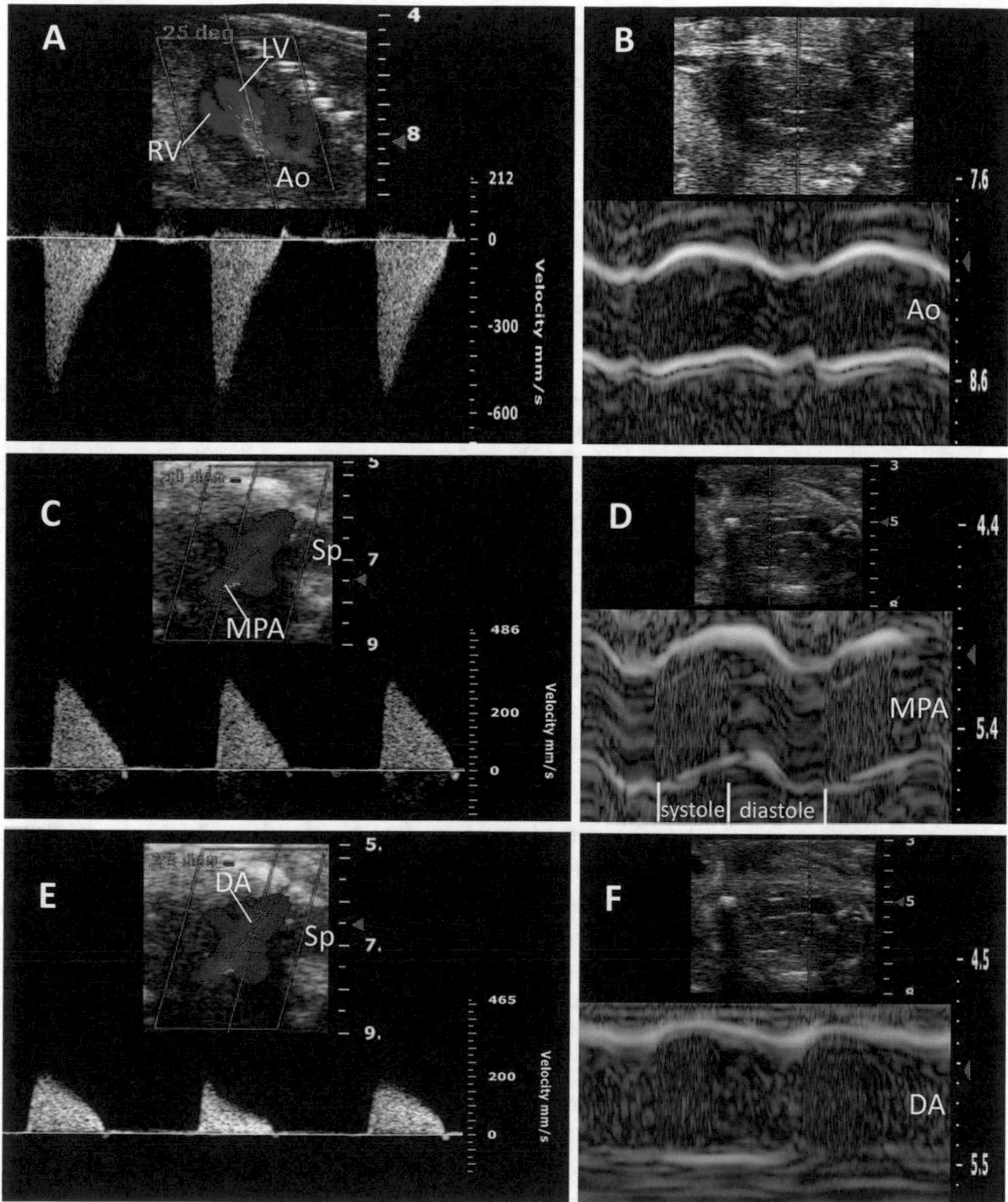

Fig. 6 Representative Doppler flow waveforms and M-mode traces of the aorta (Ao), main pulmonary artery (MPA) and ductus arteriosus (DA) of a fetus at *E*17.5. (**a**) A coronal imaging section from the left side of fetus visualizing the blood flow (in blue) from the left ventricular outflow tract to the Ao, and the Doppler velocity spectrum recorded with the sample volume at Ao. (**b**) The M-mode trace of Ao for diameter measurement. (**c**) A transverse imaging section from the left side of fetus visualizing the blood flow (in red) in MPA and DA, and the Doppler velocity spectrum recorded with the sample volume at MPA. (**d**) The M-mode trace of MPA for diameter measurement. (**e**) The pulsed Doppler flow spectrum from DA. (**f**) The M-mode recording of DA for diameter measurement. Note the change in the speckle pattern of blood flow throughout cardiac cycle differentiating the systole and diastole in the M-mode recordings of all vessels. See previous figures for other abbreviations

where the velocity waveform was measured, and an M-mode trace of the aortic walls over a cardiac cycle is recorded (Fig. 6b) (*see* **Note 8**).

5. Main pulmonary artery and ductus arteriosus: This is another important great artery to be imaged for calculation of the right ventricular outflow as well as the flow to the fetal pulmonary circulation. A transverse section at the middle level of the fetal chest is applied to visualize the flow channel between the right ventricular outflow tract and the descending thoracic aorta. Doppler color flow mapping is necessary to visualize the blood flow from the right ventricle through the main pulmonary artery and ductus arteriosus to the thoracic aorta (Fig. 6c, e). For measuring the flow in the main pulmonary artery, that is the right ventricular output, the Doppler sample volume is placed at the flow channel close to the right heart but just beyond the pulmonary valve and with intercept angle smaller than 60°. Following the main pulmonary artery downstream to the ductus arteriosus, the segment enters the thoracic aorta on the left-anterior side of the spine. The flow velocity is measured at the distal segment of the ductus arteriosus, avoiding the flow signal from the thoracic aorta. Then the transducer orientation is adjusted to obtain a perpendicular interception between the ultrasound beam and the main pulmonary artery as well as the ductus arteriosus within the current imaging section. With the vascular wall clearly visualized, the M-mode cursor line is placed at the main pulmonary artery and ductus arteriosus segments to measure their dimension changes over a cardiac cycle, respectively (Fig. 6d, f) (*see* **Notes 8**).
6. Thoracic aorta: The transducer is reorientated to get a sagittal section of the fetus. With the descending thoracic aorta in its longitudinal axis visualized on the dorsal side of the fetus, M-mode recording for diameter and pulsed Doppler for velocity are conducted at the same location at the middle segment of the thoracic aorta.
7. Common carotid artery: Also in a sagittal section of the fetus but slightly moved to one side from the midline, either the left or right common carotid artery in its long axis is visualized, and the M-mode recording for diameter and pulsed Doppler for velocity are conducted at the same location.
8. Umbilical vein and ductus venosus: With the fetal liver visualized in an oblique transverse section, the umbilical vein (intrahepatic segment) can be seen running from the surface of the fetal abdomen through the liver to join the portal venous system, with a connection to the inferior vena cava via the ductus venosus. The diameter and flow velocity of the umbilical vein are recorded at the initial segment which is straight and consistent in dimension before its slight enlargement (portal sinus)

prior to the ductus venosus. The ductus venosus, as the first shunt in the fetal circulation which connects the umbilical vein to the inferior vena cava and bypasses the liver, has the highest velocity in this region as shown by Doppler color flow mapping and a typical Doppler waveform. Its diameter and velocity are measured for flow calculation [24].

9. Three central veins: The inferior vena cava, right superior vena cava, and left superior vena cava are visualized in one single coronal imaging section from the lateral aspect of fetus. For flow calculation, the diameter and velocity are measured from the straight segment of these three vessels prior to their entrances into the right atrium.

3.4 Post-processing of Data

1. The duration of each imaging session depends on the choice of parameters to be measured, and 1 h is usually sufficient for a comprehensive evaluation of one fetus.
2. While the shape and instantaneous velocity amplitude of Doppler waveforms are used in physiological assessment, this protocol emphasizes the quantification of flow. Intensity-weighted mean velocity is traced and the corresponding velocity-time integral of the waveforms over an entire cardiac cycle are measured (*see* **Note 9**). Any retrograde flow velocity-time integral in the veins is subtracted from the antegrade flow.
3. The vascular diameter is measured from the M-mode recording. For the ascending aorta and the main pulmonary artery, the peak systolic diameter is used for flow calculation because flow ejection occurs only during systole. For other peripheral arteries with significant diastolic forward flow such as the ductus arteriosus, thoracic aorta and common carotid artery, both the peak systolic and middle diastolic diameters are measured and averaged. For veins, three diameter measurements are made at various time points throughout the cardiac cycle and averaged. The cross-sectional area of the vessel is calculated from the measured diameter by assuming a circular shape of all vessels.
4. The measurements from three cardiac cycles are averaged for both velocity-time integral and vascular diameter. The flow per cardiac cycle is calculated by multiplying the velocity-time integral by vessel area, and the flow rate (ml/min) is derived by multiplying the flow per cardiac cycle by fetal heart rate. The combined cardiac output is the sum of the flows in both the ascending aorta and main pulmonary artery.
5. The flow in each individual vessel is converted to the percentage of the combined cardiac output. Pulmonary blood flow is calculated from the difference between the main pulmonary artery and ductus arteriosus flows. The flow across the foramen ovale is estimated as the difference between the ascending aorta and pulmonary blood flow.

4 Notes

1. Mice have large litter sizes (ten embryos/fetuses or more in some strains) in each pregnancy, and an important and also challenging task is to match the phenotypes to genotypes for individual fetuses. Previously an invasive procedure with exteriorized uterine horn was used for studying the development of the fetal heart [19]. In that situation, it was feasible to identify the genotype–phenotype relationship for the fetus of interest. However, to avoid the physiological interference caused by such an invasive procedure, a noninvasive imaging procedure with intact abdomen is preferred. At a selected time point of gestation, the studied embryo/fetus can be dissected immediately after ultrasound imaging in order to match the cardiovascular phenotypes (structural and functional abnormalities) with genotypes. For a longitudinal study throughout gestation, the embryo/fetus of interest can be followed-up in several ways: (a) Using the urinary bladder as a landmark for identifying the embryo/fetus in the lower abdomen; (b) Mapping fetuses using ultrasound imaging: A recent study achieved 100% accuracy in localizing individual embryos over the gestational period using ultrasound mapping as compared with a laparotomy map, allowing genotype–phenotype correlation [27]. In some wild-type (C57BL/6J) and mutant mouse strains with smaller litter sizes (5–6 embryos/fetuses), it would be relatively easier to map and follow up the embryos/fetuses over time.
2. With many embryos/fetuses available in each litter, it would be more practical to image a limited number of embryos/fetuses, especially for a comprehensive physiological evaluation. In our experience, imaging 2–4 embryos/fetuses is possible in an imaging session of about 2 h. In a litter with uniform genotypes for all embryos/fetuses, it would be relatively easy to select a few embryos/fetuses with the most favorable locations and orientations for imaging. However, in a litter with heterogeneous genotypes in a mutant mouse strain, more effort would be needed to identify embryos/fetuses with specific abnormalities using ultrasound imaging.
3. Identification of the embryonic/fetal orientation and the corresponding anatomical appearance on ultrasound image is a very important step. Scan over the selected embryo/fetus to identify its cranial and caudal ends and the longitudinal axis of the embryonic/fetal body based on the changing body structures with the moving imaging section. Determine the left and right sides of the fetus according to the location of the fetal stomach during late gestation. With the embryonic/fetal orientation inside the maternal abdomen identified, the operator

then needs to match the fetal structures to the images on the screen of the ultrasound system. With sufficient practice and a clear 3D geometry of fetal anatomy in operator's mind, it is not difficult to perform such measurements in embryos/fetuses with less optimal orientation by modifying the imaging sections accordingly.

4. Embryonic red blood cells are large and nucleated in middle gestation but decrease in size and become enucleated as they mature. Correspondingly the echogenicity of embryonic blood under high frequency ultrasound imaging peaks at *E*13.5 and then decreases progressively toward term [28]. Thanks to this feature, real-time B-mode imaging can visualize the movement of embryonic blood cells and demonstrate the trace of the flow stream inside the heart and vessels. Compared to the hyper-echogenic blood flow, the cardiac structures such as ventricular walls, valves, and the vessel walls are relatively darker, making these soft tissue structures clearly delineated [19]. Moreover, on M-mode recording of ventricles and arteries, the echogenic blood flow inside the chamber/lumen shows sharp changes in speckle pattern throughout the cardiac cycle due to the dramatic changes in velocity of the blood cells. This feature helps pinpoint the starts of systole and diastole when embryonic/fetal ECG is not available (Figs. 4d and 6b, d, f) [24].

5. Doppler color flow mapping is essential in visualizing embryonic/fetal heart and vessels and guiding pulsed Doppler velocity measurements. The embryonic/fetal vessels are less clear in 2D imaging because of their small size, and delicate tissues with weak ultrasound reflection especially when the vessel wall should have a sharp intercept angle with ultrasound beam for velocity measurement. In addition, these vessels are in close proximity to each other. All these factors present great difficulties in differentiating individual vessels by 2D tissue imaging alone. With Doppler color flow mapping at proper settings, the blood flow in fetal vessels is able to be properly visualized, with the flow direction indicated by color. As a result, the individual vessels are easily identified and their velocity measured with angle correction. In Doppler color flow mapping, the velocity scale (pulsed repetition frequency) setting is crucial to an optimal display of flow signal. When the velocity scale is set too high, low velocity flow may not be displayed. Therefore, for embryonic heart and vessels with low velocity during middle gestation, the pulse repetition frequency should be relatively low (1–5 kHz). Conversely, when the velocity scale is set too low for the flow with relatively higher velocity, aliasing occurs in Doppler color flow mapping. The high blood flow velocity generates a Doppler shift above the Nyquist limit (a half of the pulse repetition frequency), and is displayed in

the same color as for the flow in opposite direction. However, this artifact is actually useful for quickly identifying abnormally high velocity (such as stenotic or regurgitation jet) caused by the malformation of the heart or vessels in mutant embryos/ fetuses. For a proper gain in color Doppler mapping, it is appropriate to turn the gain up until noise is encountered and then back off until the noise just clears from the image. The wall filter level is designed to remove unwanted low-frequency Doppler signals originating from slowly moving soft tissue, and the cut-off frequency is operator selectable. For imaging mouse embryonic/fetal vessels, the filter level can be gradually increased to eliminate the noise caused by tissue movement, until the color flow streams in individual vessels appear separately.

6. M-mode recording is used to measure the ventricular dimensions and systolic function after the completion of ventricular septation on *E*13.5. Although it is feasible to measure the thickness of the interventricular septum, it is challenging to measure the thicknesses of the left and right ventricular free walls. Embryonic/fetal ventricular wall has a more trabecular structure but relatively thin compact myocardial layer, and the endocardium is usually difficult to define. In such a situation, it is more practical to measure the dimension from the border of the interventricular septum to the epicardium of the ventricle at peak systole and end-diastole, and then calculate the fractional shortening as a measure of ventricular systolic function (Fig. 4c, d).
7. Pulsed Doppler waveforms from the common atrioventricular canal of the early developing heart or the mitral and tricuspid orifices of the relatively mature heart can be used for evaluating ventricular diastolic function. The *E* wave is the early diastolic filling wave caused by ventricular active relaxation, and the *A* wave is the late diastolic filling caused by the atrial contraction. The ratio of peak *E* and *A* waveforms is a commonly used measure of ventricular diastolic function. From the primitive heart tube during middle gestation to the mature heart at late gestation, we consistently observe relatively lower *E* wave and dominantly higher *A* wave, suggesting that the atrial contraction, rather than ventricular active relaxation, is the major force for ventricular filling during the embryonic/fetal period, in contrast with the mitral inflow pattern in postnatal mice. The related physiological significance is still unclear. Another important feature is that the diastolic inflow patterns are symmetrical between left and right ventricles during late gestation to term, being consistent with the symmetrical geometry of the two sides of the heart. However, both the inflow pattern and the geometry of two ventricles will become asymmetrical soon after birth [25].

8. M-mode recording provides the most accurate diameter measurement of embryonic/fetal vessels. Although it is technically challenging to manipulate the transducer to achieve a perpendicular angle between the ultrasound beam and the targeted vessel, perpendicular interception allows for the thin and delicate vessel walls to be more clearly visualized in both 2D and M-mode images because of the stronger ultrasound reflection from the perpendicular tissue interface. In addition, M-mode recording has another significant advantage in visualizing the dynamic changes of vascular dimension, ensuring a reliable flow calculation by averaging diameter measurements over a cardiac cycle [24].
9. For the flow calculation of selected major vessels throughout the fetal circulatory system, the intensity-weighted mean velocity tracing of the Doppler spectrum is used as usually done in humans [29, 30]. In some previous studies, by assuming a parabolic velocity profile across the vessel lumen, the velocity measured using maximal velocity tracing was multiplied by 0.5 and used as cross-sectional mean in the flow calculation [31]. In our studies of the mouse fetus, Doppler color flow mapping demonstrates parabolic velocity profiles in great arteries such as the aorta and main pulmonary artery, more flat or bluntly parabolic velocity distribution in major veins, and an undefinable flow profile in small flow channel like the ductus venosus. To be consistent for different kind of vessels, pulsed Doppler sample volume is adjusted to cover the whole lumen for recording flow spectrum and the intensity-weighted mean velocity of the Doppler waveform measured. A high-quality Doppler spectrum with the strongest signal, highest amplitude and sharp contours ensures the accuracy of the intensity-weighted mean velocity tracing and the subsequent flow calculation.

Acknowledgment

This work was supported by the Canadian Institutes of Health Research Grant MOP231389.

References

1. Kaufman MH (1992) The atlas of mouse development. Academic, New York, NY
2. Savolainen SM, Foley JF, Elmore SA (2009) Histology atlas of the developing mouse heart with emphasis on E11.5 to E18.5. Toxicol Pathol 37:395–414
3. Bruneau BG (2003) The developing heart and congenital heart defects: a make or break situation. Clin Genet 63:252–261
4. Teichert AM, Scott JA, Robb GB, Zhou YQ, Zhu SN, Lem M, Keightley A, Steer BM, Schuh AC, Adamson SL, Cybulsky MI, Marsden PA (2008) Endothelial nitric oxide synthase gene expression during murine embryogenesis: commencement of expression in the embryo occurs with the establishment of a unidirectional circulatory system. Circ Res 103:24–33

5. Conway SJ, Kruzynska-Frejtag A, Kneer PL, Machnicki M, Koushik SV (2003) What cardiovascular defect does my prenatal mouse mutant have, and why? Genesis 35:1–21
6. Cahill LS, Zhou YQ, Seed M, Macgowan CK, Sled JG (2014) Brain sparing in fetal mice: BOLD MRI and Doppler ultrasound show blood redistribution during hypoxia. J Cereb Blood Flow Metab 34:1082–1088
7. Cohn HE, Sacks EJ, Heymann MA, Rudolph AM (1974) Cardiovascular responses to hypoxemia and acidemia in fetal lambs. Am J Obstet Gynecol 120:817–824
8. Reuss ML, Rudolph AM (1980) Distribution and recirculation of umbilical and systemic venous blood flow in fetal lambs during hypoxia. J Dev Physiol 2:71–84
9. Di Renzo GC, Luzi G, Cucchia GC, Caserta G, Fusaro P, Perdikaris A, Cosmi EV (1992) The role of Doppler technology in the evaluation of fetal hypoxia. Early Hum Dev 29:259–267
10. Kiserud T, Kessler J, Ebbing C, Rasmussen S (2006) Ductus venosus shunting in growth-restricted fetuses and the effect of umbilical circulatory compromise. Ultrasound Obstet Gynecol 28:143–149
11. Kulandavelu S, Whiteley KJ, Bainbridge SA, Qu D, Adamson SL (2013) Endothelial NO synthase augments fetoplacental blood flow, placental vascularization, and fetal growth in mice. Hypertension 61:259–266
12. Kusinski LC, Stanley JL, Dilworth MR, Hirt CJ, Andersson IJ, Renshall LJ, Baker BC, Baker PN, Sibley CP, Wareing M, Glazier JD (2012) eNOS knockout mouse as a model of fetal growth restriction with an impaired uterine artery function and placental transport phenotype. Am J Physiol Regul Integr Comp Physiol 303:R86–R93
13. Patterson AJ, Zhang L (2010) Hypoxia and fetal heart development. Curr Mol Med 10:653–666
14. Ream MA, Chandra R, Peavey M, Ray AM, Roffler-Tarlov S, Kim HG, Wetsel WC, Rockman HA, Chikaraishi DM (2008) High oxygen prevents fetal lethality due to lack of catecholamines. Am J Physiol Regul Integr Comp Physiol 295:R942–R953
15. Seller MJ, Bnait KS (1995) Effects of tobacco smoke inhalation on the developing mouse embryo and fetus. Reprod Toxicol 9: 449–459
16. Momoi N, Tinney JP, Liu LJ, Elshershari H, Hoffmann PJ, Ralphe JC, Keller BB, Tobita K (2008) Modest maternal caffeine exposure affects developing embryonic cardiovascular function and growth. Am J Physiol Heart Circ Physiol 294:H2248–H2256
17. Bake S, Tingling JD, Miranda RC (2012) Ethanol exposure during pregnancy persistently attenuates cranially directed blood flow in the developing fetus: evidence from ultrasound imaging in a murine second trimester equivalent model. Alcohol Clin Exp Res 36: 748–758
18. Srinivasan S, Baldwin HS, Aristizabal O, Kwee L, Labow M, Artman M, Turnbull DH (1998) Noninvasive, in utero imaging of mouse embryonic heart development with 40-MHz echocardiography. Circulation 98:912–918
19. Zhou YQ, Foster FS, Qu DW, Zhang M, Harasiewicz KA, Adamson SL (2002) Applications for multifrequency ultrasound biomicroscopy in mice from implantation to adulthood. Physiol Genomics 10:113–126
20. Phoon CK, Ji RP, Aristizábal O, Worrad DM, Zhou B, Baldwin HS, Turnbull DH (2004) Embryonic heart failure in NFATc1−/− mice: novel mechanistic insights from in utero ultrasound biomicroscopy. Circ Res 95:92–99
21. Spurney CF, Lo CW, Leatherbury L (2006) Fetal mouse imaging using echocardiography: a review of current technology. Echocardiography 23:891–899
22. Hernandez-Andrade E, Ahn H, Szalai G, Korzeniewski SJ, Wang B, King M, Chaiworapongsa T, Than NG, Romero R (2014) Evaluation of utero-placental and fetal hemodynamic parameters throughout gestation in pregnant mice using high-frequency ultrasound. Ultrasound Med Biol 40:351–360
23. Foster FS, Mehi J, Lukacs M, Hirson D, White C, Chaggares C, Needles A (2009) A new 15–50MHz array-based micro-ultrasound scanner for preclinical imaging. Ultrasound Med Biol 35:1700–1708
24. Zhou YQ, Cahill LS, Wong MD, Seed M, Macgowan CK, Sled JG (2014) Assessment of flow distribution in the mouse fetal circulation at late gestation by high-frequency Doppler ultrasound. Physiol Genomics 46:602–614
25. Zhou YQ, Foster FS, Parkes R, Adamson SL (2003) Developmental changes in left and right ventricular diastolic filling patterns in mice. Am J Physiol Heart Circ Physiol 285:H1563–H1575
26. Yu Q, Shen Y, Chatterjee B, Siegfried BH, Leatherbury L, Rosenthal J, Lucas JF, Wessels A, Spurney CF, Wu YJ, Kirby ML, Svenson K, Lo CW (2004) ENU induced mutations causing congenital cardiovascular anomalies. Development 131:6211–6223
27. Ji RP, Phoon CK (2005) Noninvasive localization of nuclear factor of activated T cells c1−/− mouse embryos by ultrasound

biomicroscopy-Doppler allows genotype-phenotype correlation. J Am Soc Echocardiogr 18:1415–1421

28. Le Floc'h J, Chérin E, Zhang MY, Akirav C, Adamson SL, Vray D, Foster FS (2004) Developmental changes in integrated ultrasound backscatter from embryonic blood in vivo in mice at high US frequency. Ultrasound Med Biol 30:1307–1319
29. Hecher K, Campbell S, Doyle P, Harrington K, Nicolaides K (1995) Assessment of fetal compromise by Doppler ultrasound investigation of the fetal circulation. Arterial, intracardiac, and venous blood flow velocity studies. Circulation 91:129–138
30. Tchirikov M, Eisermann K, Rybakowski C, Schröder HJ (1998) Doppler ultrasound evaluation of ductus venosus blood flow during acute hypoxemia in fetal lambs. Ultrasound Obstet Gynecol 11:426–431
31. Vimpeli T, Huhtala H, Wilsgaard T, Acharya G (2009) Fetal cardiac output and its distribution to the placenta at 11–20 weeks of gestation. Ultrasound Obstet Gynecol 33:265–271

Check for updates

Chapter 4

Dynamic Imaging of Mouse Embryos and Cardiodynamics in Static Culture

Andrew L. Lopez III and Irina V. Larina

Abstract

The heart is a dynamic organ that quickly undergoes morphological and mechanical changes through early embryonic development. Characterizing these early moments is important for our understanding of proper embryonic development and the treatment of heart disease. Traditionally, tomographic imaging modalities and fluorescence-based microscopy are excellent approaches to visualize structural features and gene expression patterns, respectively, and connect aberrant gene programs to pathological phenotypes. However, these approaches usually require static samples or fluorescent markers, which can limit how much information we can derive from the dynamic and mechanical changes that regulate heart development. Optical coherence tomography (OCT) is unique in this circumstance because it allows for the acquisition of three-dimensional structural and four-dimensional (3D + time) functional images of living mouse embryos without fixation or contrast reagents. In this chapter, we focus on how OCT can visualize heart morphology at different stages of development and provide cardiodynamic information to reveal mechanical properties of the developing heart.

Key words Optical coherence tomography, Cardiovascular development, Embryo culture, Heart morphogenesis, Cardiodynamic analysis, Mouse, Live imaging

1 Introduction

The vertebrate embryonic heart is the first functional organ in development, and supports the growing nutritive and respiratory demands of the developing embryo. To sustain development, the early heart executes quick morphological and mechanical changes to achieve proper configuration, increase cardiac output, and maintain respiratory homeostasis. Failure to achieve these changes can have many developmental implications ranging from embryonic death to congenital heart disease, which can mildly or severely impact post utero development and the ability to thrive. Investigating the regulatory mechanisms involved in this process is essential to understand correct cardiovascular development and generate therapies to treat heart abnormalities.

Paul Delgado-Olguin (ed.), *Mouse Embryogenesis: Methods and Protocols*, Methods in Molecular Biology, vol. 1752, https://doi.org/10.1007/978-1-4939-7714-7_4, © Springer Science+Business Media, LLC, part of Springer Nature 2018

Our current understanding of embryonic cardiovascular development heavily relies on molecular analysis, histology, and tomographic and fluorescence-based imaging. These approaches have revealed valuable information in regard to gene programs and connecting them to normal developmental events or assigning them to aberrant phenotypes. While powerful, these approaches lack functional in vivo analysis to understand how biomechanics influence proper cardiogenesis. To this end, live embryo culture methods coupled with the use of optical coherence tomography (OCT) and advanced postprocessing algorithms can reveal information about the mechanical environment in the embryonic heart [1, 2].

In this chapter, we use the mouse as a model for mammalian embryonic cardiovascular development. We describe live embryo culture methods for OCT imaging to analyze live embryos, assess cardiodynamics, and quantify hemodynamics in the developing heart.

2 Materials

2.1 Embryo Dissection Medium

1. DMEM/F12: 2.5 mM L-glutamine, 15 mM HEPES buffer.
2. Fetal bovine serum (FBS), stored in 5 ml aliquots at −30 °C.
3. Penicillin/streptomycin, 100×, stored as 5 ml aliquots at −30 °C.

Combine components to obtain an 89% DMEM/F-12, 10%FBS, and 1% 100× penicillin–streptomycin solution.

2.2 Rat Serum Collection

1. Sprague Dawley male adult rats.
2. Ether.
3. Vacutainer blood collection tubes.
4. Vacutainer blood collection sets.
5. 0.45 μm syringe filter.

2.3 Embryo Dissection

1. Two #5 microdissecting forceps.
2. Microdissecting scissors.
3. Microdissecting tweezers.
4. 35 mm petri dishes.

2.4 Embryo Culture Medium

1. Embryo dissection medium as described in Subheading 2.1.
2. Rat serum, stored as 1 ml aliquots at −80 °C.

3 Methods

3.1 Rat-Serum Extraction (See Note 1)

1. Anesthetize rat with ether (*see* **Note 2**).
2. Expose dorsal aorta by abdominal incision with scissors.
3. Collect blood into a Vacutainer blood collection tube using the Vacutainer blood collection set.
4. Once blood is collected, invert the tube a few times to mix and keep it on ice while extracting blood from other rats.
5. Centrifuge tubes with collected blood at 1300 × *g* for 20 min. The supernatant should look golden. If the supernatant looks pink, which indicates erythrocyte lysis, we recommend discarding those tubes. Pool supernatant (serum) into 15 ml centrifuge tubes.
6. Centrifuge serum at 1300 × *g* for 10 min to pellet down the remaining erythrocytes. Pool serum into 50 ml centrifuge tubes.
7. Heat-inactivate serum at 56 °C for 30 min in a water bath. Keep tube lid unscrewed for ether evaporation.
8. Keep serum overnight at 4 °C with the lid unscrewed for further ether evaporation.
9. Filter serum through a 0.45 μm syringe filter.
10. Aliquot serum by 1 ml and keep at −80 °C.

3.2 Embryo Dissections

1. Set up matings and check for vaginal plugs every morning. If the plug is found, the pregnancy is taken as E0.5.
2. On the appropriate embryonic day, prepare embryo dissection medium fresh. Approximately 100 ml of medium will be used per litter.
3. Preheat dissection medium to 37 °C.
4. Preheat the dissection station and maintain at 37 °C. We use a custom-built heating station (Fig. 1a) composed of a plexiglass chamber enclosing a dissection microscope, a commercial space heater, and a temperature controller.
5. Sacrifice the pregnant mouse in accordance with established guidelines.
6. Dissect the uterus with embryos and transfer over to warm (37 °C) dissection medium in a petri dish. Immediately move the dish into the temperature controlled dissection station.
7. With micro dissecting forceps, under microscope guidance, sequentially separate individual embryos from the muscular uterine tissue.
8. Delicately peel away the deciduum covering the embryo. Leave large piece of the deciduum connected to the ectoplacental cone. This remaining tissue can be shaped with

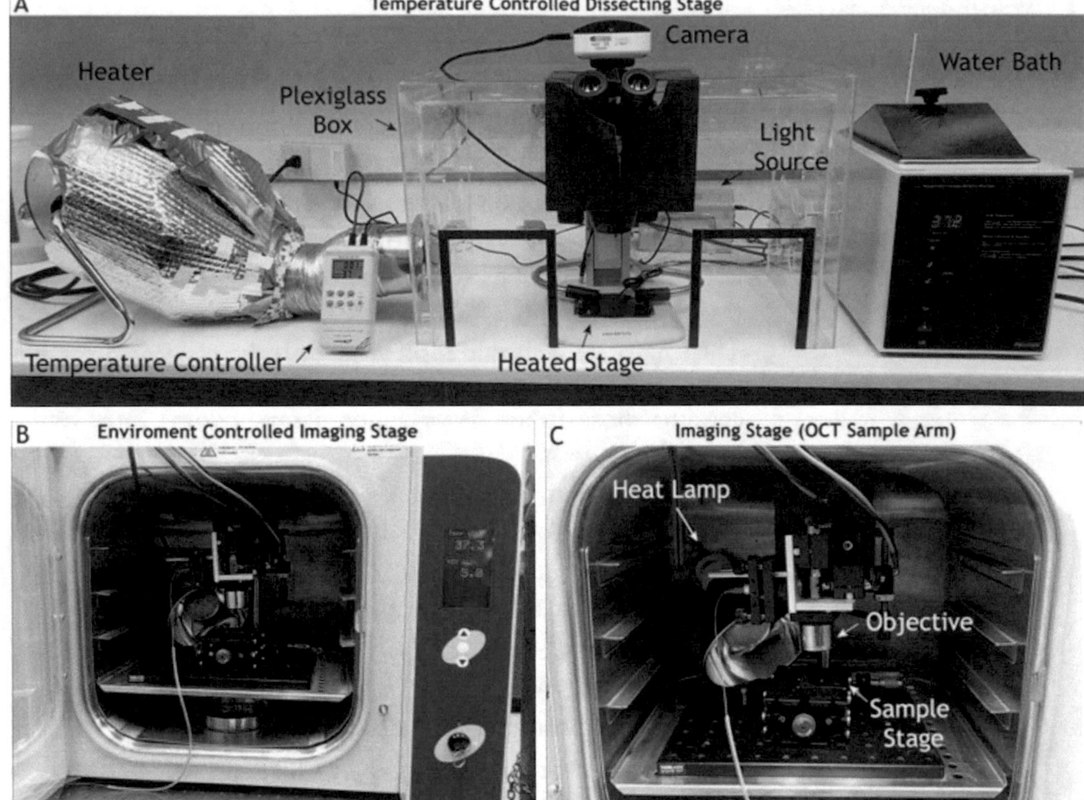

Fig. 1 Dissection and imaging stage. (**a**) Embryos are dissected in a heated lab-built dissection stage consisting of a plexiglass box, an Arduino controlled heated stage, and a commercial heater connected to the plexiglass box and modulated by a temperature controller. (**b**) Dissected embryos are imaged in a commercial incubator maintaining 37 °C and 5%CO_2. (**c**) OCT sample arm resides within the incubator to image live embryos. Heating lamp is used to maintain heat while incubator is open

dissection forceps to position the embryo with the heart toward the imaging beam.

9. Delicately peel away the Reichert's membrane exposing the yolk sac, while leaving the yolk sac intact (*see* **Note 3**).
10. Transfer the embryo to a culture dish with fresh warm media using a transfer pipette.
11. For short-term experiments limited to a few hours (e.g., for live analysis of cardiodynamics), embryos can be imaged in dissection media. However, for longer imaging sessions, when embryos are expected to progress through development, the culture medium should be supplemented with rat serum: two parts DMEM/F12, one part rat serum, and 1% 100× penicillin–streptomycin.
12. Place the culture dish in a 5%CO_2 and 37 °C incubator for 30 min to 1 h for the embryo to recover.

3.3 Imaging Setup

Different OCT systems can be utilized for mouse embryo imaging. To acquire images presented in this chapter, we utilized a lab-built spectral-domain OCT system based on the MICRA 5 Titanium-Sapphire laser (Coherent Inc.) with a central wavelength of 800 nm and a bandwidth of ~110 nm [1]. Fiber-based Michelson interferometer is used for the interference of light reflected from the reference arm and the back reflected light from the sample arm to form fringes. Interference fringes are spatially resolved using a lab-built spectrometer consisting of a diffraction grating, a focusing lens system, and a line-field CMOS camera (spL4096–140 km, Basler). A 1D depth profile (A-scan) is obtained by performing a fast Fourier transform on fringe data that is transformed into equally sampled data in linear k-space using the laser spectra information. 2D images and 3D volumes are obtained using galvanometer scanners (Cambridge Technology 6220H with 8 mm mirror) mounted on the sample arm over the imaging objective. The system has an A-line rate of ~68 kHz, and provides the spatial resolution of ~5 μm (in tissue) in the axial direction and ~4 μm in the transverse direction. The system sensitivity was measured at ~97 dB with a sensitivity drop of ~4 dB over ~1 mm in depth. The OCT sample arm is placed in a commercial environment-controlled incubator during the entire imaging process (Fig. 1b, c) (*see* **Notes 4** and **5**).

3.4 Visualization of Embryonic Structures Using OCT

OCT imaging of live static culture provides visualization of embryonic structural features through early development (*see* **Note 6**). While not in a natural physiological environment, embryos cultured at E7.5–E10.5 stages reach the same developmental hallmarks as embryos in utero [1–3]. Cellular level resolution allows for visualization and analysis of structural defects in embryonic mouse mutants. As an example, Fig. 2 shows structural phenotypic analysis of embryos at E8.5 and E9.5 with OCT in the Wdr19 mouse model. Wdr19 mutants are embryonic lethal at E10.5 and exhibit defects in neural tube closure and cardiac looping [1]. The figure shows bright-field microscopic images (left) next to 3D OCT reconstructions (right) of Wild Type embryos as controls (Fig. 2a, b) and Wdr19 mutant embryos (Fig. 2c, d) at two developmental stages. The structure of the head region is clearly distinct. In addition to size difference, it can be seen that the control embryo at E9.5 has a well-closed neural tube at the head region while the Wdr19 embryo exhibits an open neural tube. In addition, the structure of the heart tube is abnormal in the Wdr19 embryo at E8.5 demonstrating a deficiency in cardiac looping, which is essential for normal cardiogenesis. This defect is clearly seen in the OCT reconstruction.

For later stage embryos, the imaging depth of the OCT might be insufficient to visualize internal structures. In this case, embryos can be dissected out of the yolk sac for imaging. While this approach is not compatible with live imaging (since it dis-

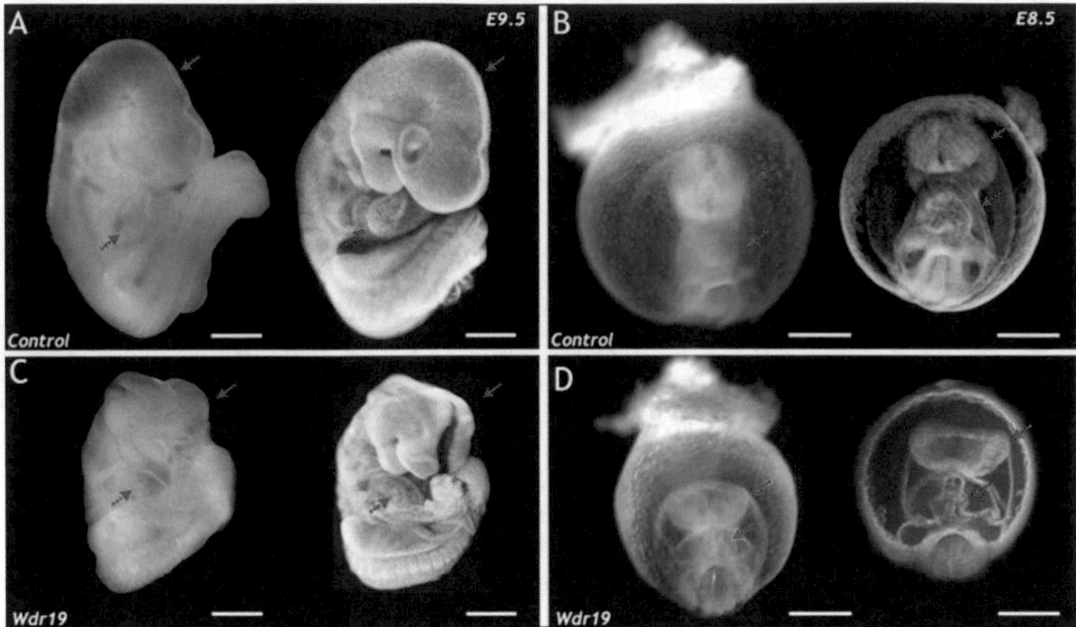

Fig. 2 Structural analysis of early stage mouse embryos. Brightfield microscopic images (left) and 3D OCT images (right) of control embryos (**a**, **b**) and Wdr19 mutant (**c**, **d**) embryos at E9.5 (**a** + **c**) and E8.5 (**b** + **d**). Solid arrows point at the neural tube in the head region and dashed arrows point at the heart. Scale bars correspond to 500 μm. Adapted from [1]

rupts circulation), it reveals structural features at cellular level without application of contrast agents in freshly dissected embryos, which might be beneficial in a variety of applications.

3.5 Cardiodynamic Analysis

A high volume acquisition rate is critical for cardiodynamic analysis. Nonetheless, traditional OCT systems cannot achieve the ~100 frames per second frame rate required for direct volumetric acquisition and heart wall tracking. To resolve this limitation, a number of algorithms have been proposed to synchronize time sequences acquired over multiple heartbeats at neighboring positions through the entire heart volume [4–9]. These methods rely on the periodic nature of the cardiac cycle. Below is an algorithm based on nongated synchronization (Fig. 3), which we successfully used for reconstruction of 4D beating hearts [1, 3]. This approach is best for embryos from E8.5 to E9.5.

1. Position the embryo in the culture dish so that the heart region of the embryo faces up toward the imaging beam. Leaving a piece of deciduum attached to the ectoplacental cone will allow one to manipulate the tissue into a base and keep the embryo steady.
2. Once the embryo is positioned, spatial acquisition parameters have to be assigned to assure that the entire volume of heart will be imaged. For E8.5 embryos, our scanning amplitude in the *X*

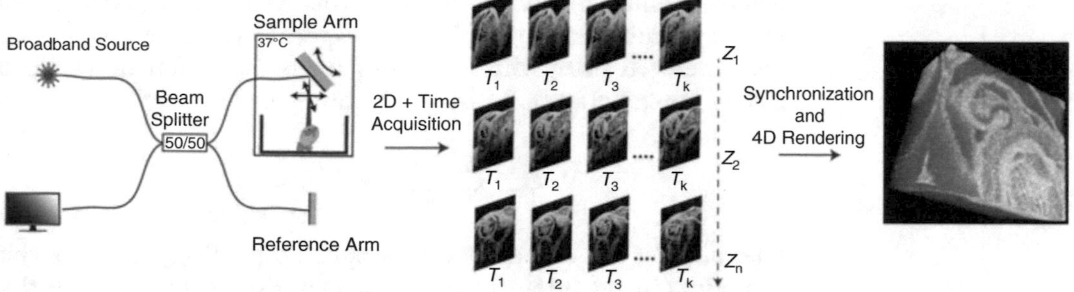

Fig. 3 Volumetric cardiodynamic imaging approach. Adapted from [1]

and *Y* direction is ~750 μm ± 150 μm. From this point on, the embryo should remain undisturbed in the incubator for the duration of the imaging session.

3. Make note of the assigned spatial parameters, as they will be necessary for computing the spatial geometry of your sample during postprocessing and rendering.
4. Assign scanning parameters. As a suggestion, we assign 600 A-lines and 20,000 B-scans per volume for an E8.5 embryo. The B-scans are continuously acquired at 100 frames per second over a single volume (the *Y*-galvanometer scanner is slowly moving through the volume while the *X*-galvanometer scanner is quickly moving for B-scan acquisition). As a result, during each heartbeat cycle (the heart rate of the embryo is ~2 Hz at this stage) the *Y* scanner moves by ~2 μm, which is below the lateral resolution of the system.
5. After imaging is completed, OCT data files must be processed into an image sequence of image files corresponding to B-scans for processing and visualization.
6. Define the number of frames per heartbeat cycle. Split the total data set into individual cycles and assign the position for each heartbeat cycle according to the total scan size in the *Y* direction and the total acquired number of heartbeats. We use Imaris v8.0.2 software (Bitplane, Switzerland) for 3D rendering and processing. The heartbeats are synchronized to the same phase of the heartbeat cycle using previously described algorithms [10], revealing the 4D (3D plus time) cardiodynamic data set.

3.6 Imaging the Vasculature Using Speckle Variance

Vascular morphology of the yolk sac and the embryo proper can be visualized using speckle variance (SV) OCT approach. This method relies on the analysis of temporal variations in pixel intensity in structural OCT images. Interaction of laser radiation with tissue scatterers creates a speckle pattern. The SV OCT method segments out varying pixel intensities due to moving scatterers at specific positions. Since most moving scatterers in embryos are blood cells, this method can be used to reveal vascular structure [11, 12]. Speckle variance images can be generated using the following formula, which determines the variance between N images in time, t:

$$\mathrm{SV}_{i,j} = \frac{1}{N}\sum_{t=0}^{N-1}\left(I_{i,j,t} - I_{i,j,\mathrm{mean}}\right)^2$$

where, $\mathrm{SV}_{i,j}$ is the variance at a specific pixel (i,j), $I_{i,j,t}$ is the structural image intensity at that pixel at time t, and $I_{i,j,\mathrm{mean}}$ is the mean intensity of pixel (i,j) for all N values [11]. Speckle variance approach provides vascular structure in the developing embryo and extra-embryonic structures; however, movements associated with the beating embryonic heart can degrade the SV signal, and therefore, limits it to nonmoving vascular beds.

3.7 Live Imaging of Hemodynamics Using Doppler OCT

Quantitative blood flow analysis is an important component to functionally assess the heart. Doppler OCT relies on the measurement of OCT phase shifts between consecutive A-scans at each pixel and can be used to quantify hemodynamics at the same spatial and temporal resolution as structural OCT imaging [13]. The direction of flow relative to the laser beam needs to be defined for quantitative blood flow analysis. The flow velocity at each pixel can be reconstructed using the formula [14]:

$$v = \frac{\Delta\varphi}{2n\dfrac{2\pi}{\lambda}\tau\cos\beta}$$

where φ is a Doppler phase shift between adjacent A-scans, n is the refractive index, λ is the central wavelength used for imaging, τ is the time between consecutive A-scans, and β is the angle between the flow direction and the laser beam. The angle β is calculated from structural 2D and 3D data sets acquired from imaged embryos. To image embryonic mice we usually assume a refractive index of $n = 1.4$.

The resolution of Doppler OCT allows for quantitative hemodynamic analysis from bulk blood cell movement [15] as well as individual circulating erythrocytes at early circulation stages, when the majority of blood cells are still restricted from flowing in the blood islands [16]. Spatially and temporally resolved blood flow profiles can be reconstructed from these measurements. Hemodynamic measurements acquired by Doppler OCT in cultured mouse embryos are comparable to measurements made using fast-scanning confocal microscopy in superficial vessels. However, flow measurements deep within the embryo are not possible with confocal microscopy due to the imaging depth limitation of this method, and unique to the Doppler OCT technique [15, 17].

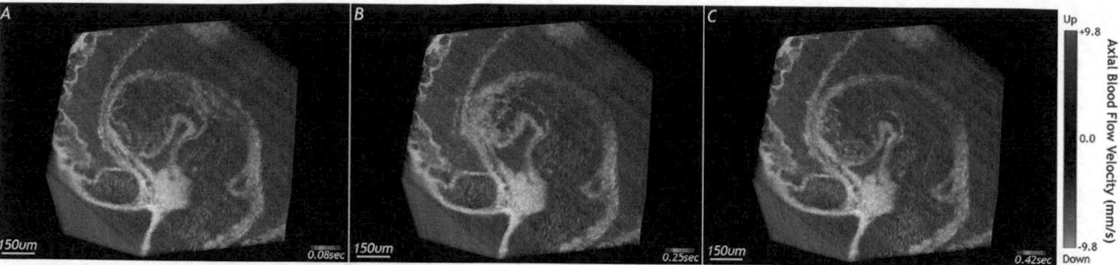

Fig. 4 Volumetric hemodynamic imaging in the embryonic heart. Snapshots from 4D hemodynamic imaging of a late E8.5 wild-type embryonic mouse heart at different phases of the cardiac cycle. The structural data is shown in gray scale while the Doppler OCT velocity measurement is overlaid in color

Volumetric blood flow analysis in the beating heart with Doppler OCT can be performed using the synchronization approach described above for structural imaging [3]. For that, structural and Doppler OCT images are exported and treated as two channels of the same data set. The synchronization is performed based on the structural data set, while the Doppler OCT frames are passively rearranged and registered based on the corresponding structural data arrangements. Figure 4 shows an example of volumetric Doppler OCT analysis of blood flow in the embryonic heart. The blood flow labeled with blue (toward the detector) and magenta (away from the detector) is clearly distinguishable in the heart and extraembryonic vessels at different phases of the heartbeat cycle.

Analysis of absolute blood flow velocity in the heart might be challenging due to difficulty of evaluating the flow direction. At early heart tube stages, when the flow is laminar and the direction of flow is obvious based on the 3D structure of the heart tube, the absolute flow velocity profiles in the heart can be reconstructed and are comparable with ultrasonic measurements acquired in utero [3]. This provides an opportunity for detailed quantitative volumetric biomechanical analysis in the embryonic heart by correlating heart wall dynamics with local flow measurement [3]. An example of correlative analysis of the cardiodynamics and hemodynamics is shown in Fig. 5. This can reveal the details of the cardiac cycle previously not known and opens a door for studies on biomechanical regulation of cardiac function.

3.8 Conclusion

Here we describe methods for dynamic imaging of the developing embryonic mouse heart. Using OCT and postprocessing methods, one can extract detailed structural and functional information. This is highly relevant when trying to understand how dynamic and mechanical changes influence development and result in mutant phenotypes. Combined with molecular genetic approaches these methods can help us generate a more complete view of all the parameters that regulate development.

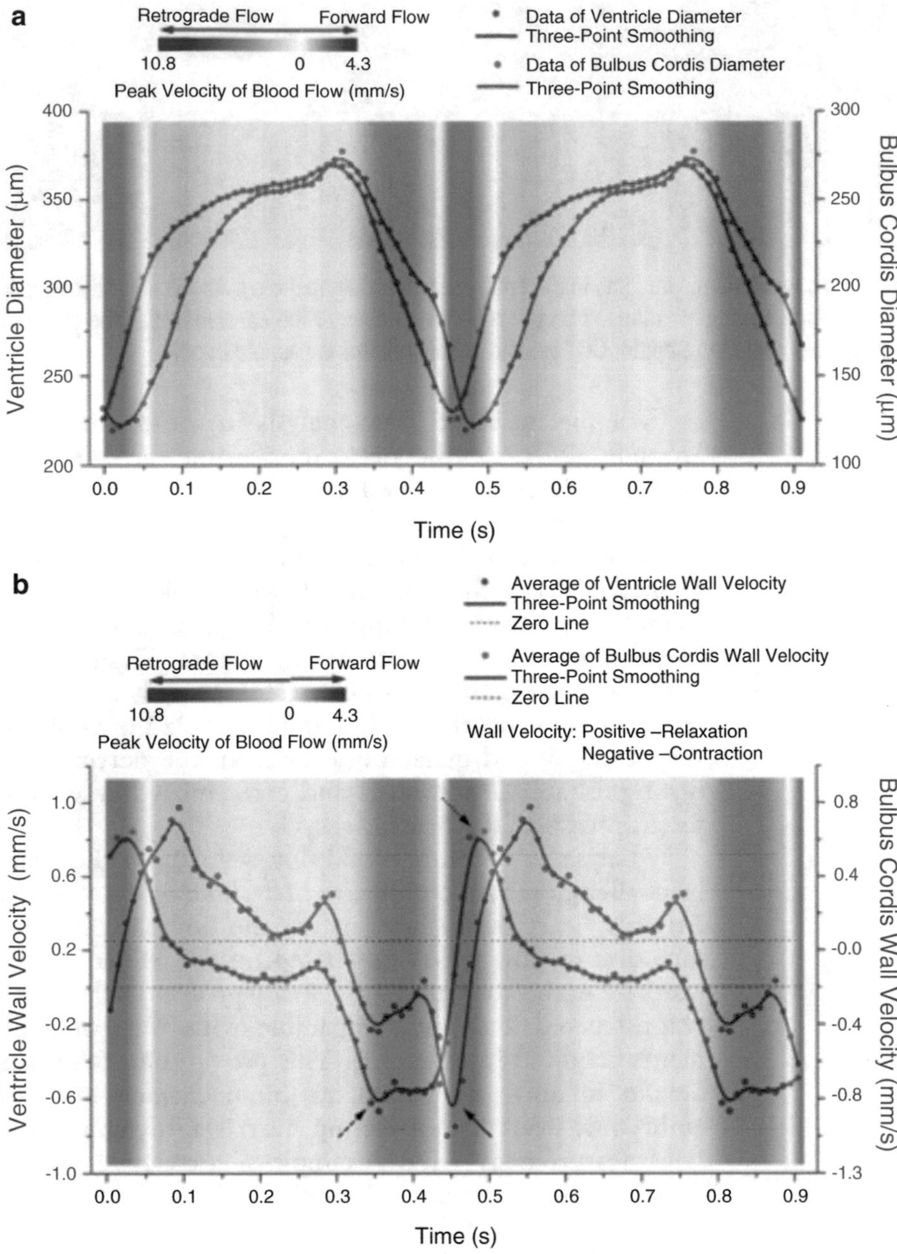

Fig. 5 Analysis of hemodynamics in relation to heart wall dynamics. (**a**) Diameters and (**b**) average heart wall velocities of the ventricle and the bulbus cordis are plotted with blood flow dynamics in the bulbo-ventricular region of a Wild-Type E9.0 embryo. The heartbeat is duplicated for a second cycle for better visualization. Adapted from [3]

4 Notes

1. Rat serum is available commercially; however, we prepare our own serum via the protocol provided here.
2. We use ether for rat serum extraction because it evaporates from the serum unlike other anesthetics.
3. We find it helpful to change out the media during the dissection to assure pH and temperature are consistent.
4. We recommend tilting the scanning head slightly (~5°) to minimize light reflection from the media surface and the dish bottom, which might significantly improve the imaging quality.
5. We recommend removing the scanning head from the incubator when not in use or turning off the incubator and removing the water tray from the stage to minimize the effect of moisture on the scanning head's electronics and mechanics.
6. It is essential to maintain 37 °C through out the imaging session for the imaging stage, the culture media, and the dissecting stage. Temperature variations can have a dramatic effect on cardiodynamics and embryo viability.

Acknowledgments

This work is supported by the National Institute of Health with grants R01HL120140, U54HG006348, R01HD086765, and T32HL07676, and by the Optical Imaging and Vital Microscopy Core at Baylor College of Medicine.

References

1. Lopez AL III, Wang S, Larin KV, Overbeek PA, Larina IV (2015) Live four-dimensional optical coherence tomography reveals embryonic cardiac phenotype in mouse mutant. J Biomed Opt 20:90501
2. Wang S, Garcia MD, Lopez AL, Overbeek PA, Larin KV et al (2017) Dynamic imaging and quantitative analysis of cranial neural tube closure in the mouse embryo using optical coherence tomography. Biomed Opt Express 8:407–419
3. Wang S, Lakomy DS, Garcia MD, Lopez AL III, Larin KV et al (2016) Four-dimensional live imaging of hemodynamics in mammalian embryonic heart with Doppler optical coherence tomography. J Biophotonics 9:837–847
4. Liu A, Wang R, Thornburg KL, Rugonyi S (2009) Efficient postacquisition synchronization of 4-D nongated cardiac images obtained from optical coherence tomography: application to 4-D reconstruction of the chick embryonic heart. J Biomed Opt 14:044020
5. Jenkins MW, Rothenberg F, Roy D, Nikolski VP, Hu Z et al (2006) 4D embryonic cardiography using gated optical coherence tomography. Opt Express 14:736–748
6. Jenkins MW, Chughtai OQ, Basavanhally AN, Watanabe M, Rollins AM (2007) In vivo gated 4D imaging of the embryonic heart using optical coherence tomography. J Biomed Opt 12:030505
7. Mariampillai A, Standish BA, Munce NR, Randall C, Liu G et al (2007) Doppler optical cardiogram gated 2D color flow imaging at 1000 fps and 4D in vivo visualization of embryonic heart at 45

fps on a swept source OCT system. Opt Express 15:1627–1638

8. Larin KV, Larina IV, Liebling M, Dickinson ME (2009) Live imaging of early developmental processes in mammalian embryos with optical coherence tomography. J Innov Opt Health Sci 2:253–259
9. Gargesha M, Jenkins MW, Wilson DL, Rollins AM (2009) High temporal resolution OCT using image-based retrospective gating. Opt Express 17:10786–10799
10. Liebling M, Forouhar AS, Gharib M, Fraser SE, Dickinson ME (2005) Four-dimensional cardiac imaging in living embryos via postacquisition synchronization of nongated slice sequences. J Biomed Opt 10:054001
11. Sudheendran N, Syed SH, Dickinson ME, Larina IV, Larin KV (2011) Speckle variance OCT imaging of the vasculature in live mammalian embryos. Laser Phys Lett 8:247–252
12. Mariampillai A, Leung MK, Jarvi M, Standish BA, Lee K et al (2010) Optimized speckle variance OCT imaging of microvasculature. Opt Lett 35:1257–1259
13. Chen Z, Milner TE, Srinivas S, Wang X, Malekafzali A et al (1997) Noninvasive imaging of in vivo blood flow velocity using optical Doppler tomography. Opt Lett 22:1119–1121
14. Vakoc B, Yun S, de Boer J, Tearney G, Bouma B (2005) Phase-resolved optical frequency domain imaging. Opt Express 13:5483–5493
15. Larina IV, Sudheendran N, Ghosn M, Jiang J, Cable A et al (2008) Live imaging of blood flow in mammalian embryos using Doppler swept-source optical coherence tomography. J Biomed Opt 13:060506
16. Larina IV, Ivers S, Syed S, Dickinson ME, Larin KV (2009) Hemodynamic measurements from individual blood cells in early mammalian embryos with Doppler swept source OCT. Opt Lett 34:986–988
17. Jones EA, Baron MH, Fraser SE, Dickinson ME (2004) Measuring hemodynamic changes during mammalian development. Am J Physiol Heart Circ Physiol 287:H1561–H1569

Check for updates

Chapter 5

In Vivo Imaging of the Mouse Reproductive Organs, Embryo Transfer, and Oviduct Cilia Dynamics Using Optical Coherence Tomography

Shang Wang and Irina V. Larina

Abstract

The oviduct (or fallopian tube) serves as the site where a number of major reproductive events occur for the start of a new life in mammals. Understanding the oviduct physiology is essential to uncover hidden mechanisms of the human reproduction and its disorders, yet the current analysis of the oviduct that is largely limited to in vitro imaging is a significant technical hurdle. To overcome this barrier, we have recently developed in vivo approaches based on optical coherence tomography for structural and functional imaging of the mouse oviduct. In this chapter, we describe the details of such live imaging methods that allow for three-dimensional visualization of the oviduct wall morphology, microscale mapping of the oviduct cilia beat frequency, and high-resolution observation of the cumulus–oocyte complex at the cellular level. We expect this set of imaging tools will enable novel studies toward a comprehensive knowledge of the mammalian reproduction.

Key words Mouse Oviduct, Mammalian Reproduction, In Vivo Imaging, Optical Coherence Tomography, Cumulus–Oocyte Complex, Motile Cilia, Cilia Beat Frequency

1 Introduction

In mammals, a life begins inside the oviduct (or fallopian tube) that transports gametes, serves as the fertilization site, and supports the early embryogenesis [1, 2]. Despite its significant role in the mammalian reproduction, the underlying mechanism of how the oviduct regulates these reproductive events remains poorly understood. A number of essential questions such as what special microenvironment the oviduct provides during the embryo transfer [3] and what exact role the cilia are playing in a successful fertilization [4] are yet to be answered. Traditional investigations are limited to in vitro imaging analysis with the techniques including scanning electron microscopy [5], confocal fluorescent microscopy [6] and bright-field microscopy [7] that lack the capability and potential to assess the oviduct tissues and cells in their natural

Paul Delgado-Olguin (ed.), *Mouse Embryogenesis: Methods and Protocols*, Methods in Molecular Biology, vol. 1752, https://doi.org/10.1007/978-1-4939-7714-7_5,

conditions. The absence of an in vivo oviduct imaging approach allowing for live studies of the mammalian reproduction is becoming a major technical hurdle which prevents our understanding of fertility, infertility, and early embryonic development.

Based on optical coherence tomography (OCT), we have recently developed methods for structural and functional imaging of the mouse oviduct in vivo [8, 9]. Taking advantage of the unique imaging scale of OCT [10] and relying on the speckles from low-coherence interferometry [11], our live imaging approach is able to provide three-dimensional (3D) reconstruction of the oviduct wall morphology, depth-resolved mapping of the oviduct cilia beat frequency (CBF), and high-contrast visualization of the cumulus–oocyte complex inside the oviduct, all with microscale spatial resolution and millimeter-level field of view. In this chapter, we will discuss this approach in detail, aiming to facilitate the understanding and the practical implementation of the methods.

In Subheading 2, we will describe the OCT system and the tools that are needed for surgery and experimental setup. The required specifications of an OCT system to fulfil the oviduct imaging purpose will be discussed in the Subheading 4. In Subheading 3, we will first describe the in vivo setup for OCT imaging of the mouse oviduct. Then the imaging methods will be detailed with two subsections, the structural imaging including oviduct wall and cumulus–oocyte complex and the functional imaging focused on mapping of CBF. For both of these aspects, the principle of imaging will be briefly described and the specific steps for data acquisition and postprocessing will be discussed. Examples of the imaging results will be given by the end of the descriptions.

2 Materials

2.1 OCT System

An OCT system, either commercial or home-built, is required. The suggested specifications of the system are discussed in **Note 1**, which can be used as a guide when purchasing or building an OCT system for in vivo imaging of the mouse oviduct. In our study, we employed a home-built spectral domain OCT system [8, 9]. The system has a broadband Ti:Sapphire laser source (Micra-5, Coherent). The wavelength of the laser light is centered at ~808 nm with a ~110 nm bandwidth. The open-space light is coupled into a single mode fiber, and a fiber-based Michelson interferometer (50:50 wideband fiber coupler, Thorlabs) is utilized for interference of the light backscattered and reflected from the sample and reference arms, respectively. In the sample arm, a set of two orthogonal mirror galvanometers are used for transverse scanning of the light, and a 5× scan lens (Thorlabs) is used to focus the light into the sample. The interference fringes are spatially resolved with a spectrometer consisting of 1800 l/mm transmission gratings

(Wasatch Photonics), a focusing lens system, and a line-field CMOS camera (spL2048-140 km, Basler). We use a frame grabber (PCIe-1433, National Instrument) to acquire data from the camera. The obtained wavelength-dependent interference raw signal is interpolated for equal *k*-space and a fast Fourier transform is performed to get the depth-resolved intensity A-line. The control of the OCT system, including scanning of light, data acquisition, and real-time visualizations of A-line and B-scan, is through a home-developed LabVIEW program. Our OCT system provides an axial resolution of ~5 μm in tissue (assuming a refractive index of 1.4) and a transverse resolution of ~4 μm. A total of 2048 pixels are available in depth with the pixel scale of ~2 μm in tissue. With the camera exposure time of 18 μs, the system has a sensitivity of ~95 dB at a ~50 μm optical path-length difference. The sensitivity drop over depth is around 4 dB per millimeter. The OCT system has an A-line rate up to ~68 kHz, and the transverse scanning field of view can reach millimeter level.

2.2 Surgery and Experimental Setup

1. Mouse anesthesia, can be either an isoflurane-based anesthesia device or a 1.25% tribromoethanol solution for intraperitoneal injection.
2. Heating pad, to maintain the body temperature at 37 °C for the anesthetized mouse.
3. Hair removal cream, to remove the hair at the surgical site.
4. 70% ethanol, for skin disinfection.
5. A pair of straight operating scissors, for cutting the skin and muscle layer.
6. A pair of blunt forceps, for grabbing the reproductive organs.
7. A pair of straight surgical clamps, for holding the ovarian fat pad.
8. Two L-shape optical table clamps, a mini lab jack, and screws, for stabilizing the surgical clamps.
9. A dissection scope, for observing and adjusting the oviduct position prior to OCT imaging.

3 Methods

3.1 In Vivo Setup for OCT Imaging

The animal manipulation to have the oviduct exposed to the OCT imaging beam is illustrated in Fig. 1. This procedure is similar to the one that is routinely performed for injecting zygotes in transgenic mouse generation [12]. Thus, the surgery is not expected to cause complications related to mouse reproduction. The surgical procedure should be conducted under a dissection microscope for better visualization and control. All surgical tools listed in Subheading 2.2 can be prewarmed (~37 °C) to minimize the temperature influence.

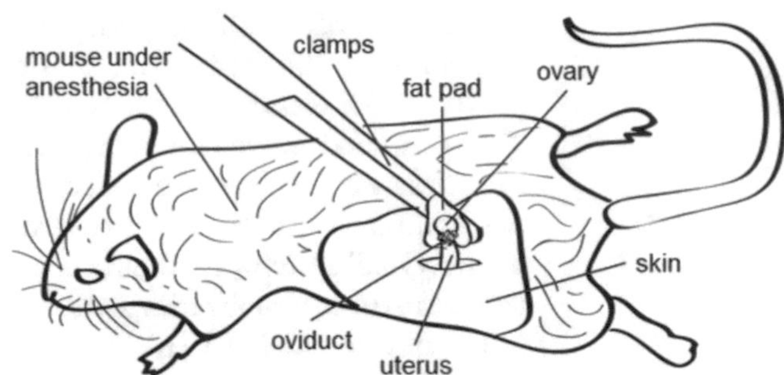

Fig. 1 Experimental setup for in vivo OCT imaging of the mouse oviduct. Reproduced from [9]

1. Put the female mouse under anesthesia using established guidelines. Depending on the available resources and animal protocols, the mouse can be anesthetized with either inhalation of isoflurane or intraperitoneal injection of 1.25% tribromoethanol solution.
2. Place the mouse on the heating pad on its belly to keep a body temperature of 37 °C during the whole experiment.
3. Use hair removal cream on one side of the dorsal midline (*see* Fig. 1) to expose the skin.
4. Swab the exposed skin with 70% ethanol.
5. Use the operating scissors to make an incision of about 1 cm long in the skin and muscle layer.
6. Locate the reproductive organs through the incision, including the ovary, oviduct and part of the uterus.
7. Use the blunt forceps to grab the ovarian fat pad and gently pull the reproductive organs out of the body cavity (*see* Fig. 1).
8. Hold the fat pad with the blunt forceps to adjust and select an angle and position for clamping. This is to ensure that the oviduct is facing up and can be easily accessed with the OCT imaging beam later during the experiment (*see* **Note 2**).
9. Clamp the fat pad with the surgical clamps (*see* Fig. 1) and fix the handle of the clamps onto the lab jack with the L-shape clamp (*see* **Note 3**).
10. Place the whole setup (together with the heating pad) under the OCT imaging beam and fix the lab jack onto the optical table with the L-shape clamp.

3.2 Structural Imaging of the Mouse Oviduct

The OCT imaging contrast is formed by the mismatch of tissue refractive indexes, which produces different amplitudes of backscattering. With OCT structural imaging, the morphology of the oviduct wall, including the dimension and the shape of the mucosal folds inside the oviduct are directly visualized in 3D at a microscale resolution. The gradual morphological change of the mucosal folds from the ampulla to the isthmus can be analyzed [8]. Given the size of the mouse oocytes and cumulus cells that are larger than the OCT spatial resolution [13], individual oocytes as well as individual cumulus cells surrounding the oocytes can be well resolved inside the lumen of the oviduct. The 3D localization of the cumulus–oocyte complex can be performed.

1. Locate the region of interest (ampulla, isthmus, or their junction) of the oviduct by observing and adjusting the position of OCT imaging beam on the tissue surface.
2. Perform 2D depth-resolved B-scan imaging with real-time visualization to optimize the depthwise location of the focal plane relative to the tissue (*see* **Note 4**).
3. Adjust the scanning distance for both the transverse directions (*see* **Note 5**) and conduct 3D OCT volumetric imaging of the tissue (*see* **Note 6**).
4. Export 2D B-scan data with logarithm scale into the format of images, such as .tiff (*see* **Note 7**).
5. Reconstruct 3D volume of the oviduct tissue with 3D software, such as Imaris (Bitplane) or ImageJ, and perform desired image visualization (e.g., transparency and cross section) and analysis. Examples of the imaging results are shown in Fig. 2a, b for the oviduct wall morphology and the cumulus–oocyte complex inside the oviduct, respectively (*see* **Note 8**).

3.3 Functional Imaging of the Oviduct Cilia

Cilia in the mouse oviduct are ~300 nm in diameter and ~5 μm in length, which does not allow a direct OCT visualization due to insufficient resolution. Our functional OCT imaging of the oviduct cilia is based on their dynamic behavior, the periodic movement. The beat of cilia leads to variation of the OCT speckles [14]. A spectral analysis of the OCT speckles over time within the B-scan produces distinct peaks where the cilia are located [15]. Thus, a mapping of the peak amplitude can reveal the cilia location inside the oviduct and a mapping of the frequency position of the peak can reveal the CBF [9]. This functional approach enables comprehensive studies of the oviduct ciliary activities in relation to physiological stages and anatomic locations [9]. The described data processing procedure below is realized using Matlab (MathWorks).

1. Select the region of interest from the mouse oviduct based on a real-time visualization of the OCT B-scan imaging of the oviduct tissue.
2. Acquire time-lapse of B-scans at the selected location (*see* **Note 9**).
3. Convert all the intensity amplitudes into logarithm scale and store the B-scan data over time into a 3D matrix.

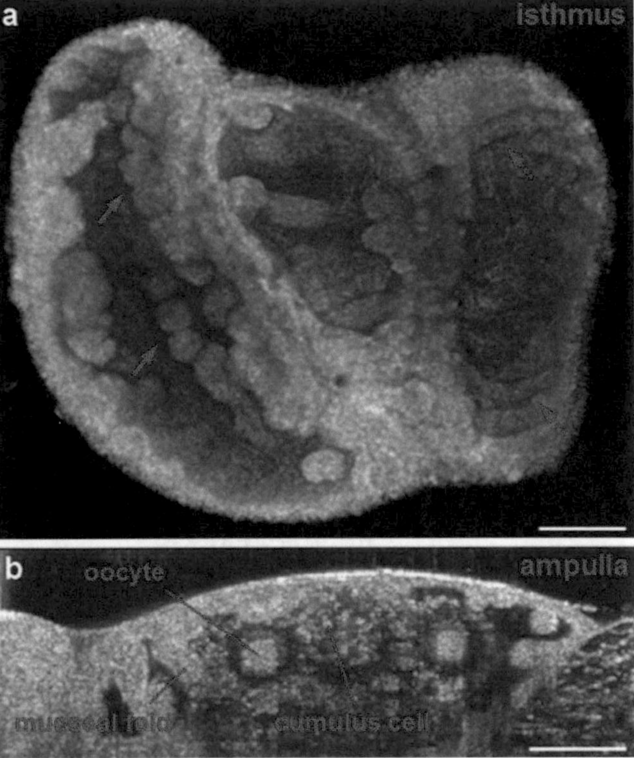

Fig. 2 Structural OCT imaging of the mouse oviduct in vivo shows (**a**) the oviduct wall morphology at isthmus and (**b**) the cumulus–oocyte complex inside the lumen of ampulla. Solid and dotted arrows in (**a**) indicate different shapes of mucosal folds at the anterior and posterior isthmus, respectively. Scale bars correspond to 200 μm. Reproduced from [8] with permission

4. Perform the fast Fourier transform to the temporal OCT data at each B-scan position to obtain the amplitude spectra (*see* **Note 10**).
5. Create a binary mask based on the averaged B-scan OCT data over time to eliminate the background.
6. Plot together all the amplitude spectra profiles associated with the oviduct tissue to set the threshold for determining whether a peak in the spectral domain is related to the beat of cilia (*see* **Note 11**).
7. Map the peak amplitudes that are beyond the threshold to the structural B-scan image to show the cilia location in the depth-resolved 2D field of view.
8. Generate a binary mask highlighting the location of cilia.
9. Determine the thresholds of frequency for CBF mapping (*see* **Note 12**).

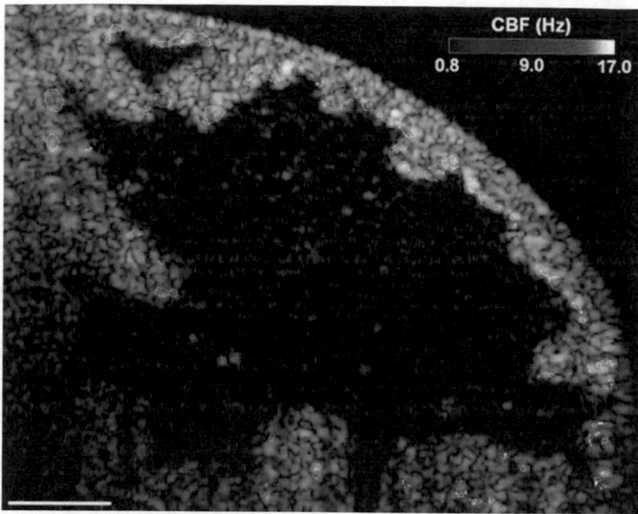

Fig. 3 Functional OCT imaging of the mouse oviduct in vivo shows the mapping of oviduct CBF in a depth-resolved field of view at the junction of ampulla and isthmus. Scale bar corresponds to 100 μm. Reproduced from [9]

10. Map the peak position within the thresholds to the structural B-scan image as the final depth-resolved 2D imaging of the oviduct CBF. An imaging result of the CBF at the junction of the ampulla and isthmus is shown as an example in Fig. 3.

4 Notes

1. The requirement of the system specifications comes from three aspects: spatial resolution, sensitivity, and imaging speed. Although a microscopy-level spatial resolution (e.g., 1–2 μm) is not required, higher resolution greatly benefits the imaging of fine structures. For both structural and functional imaging of the mouse oviduct, we recommend a spatial resolution of at least 5–6 μm in tissue with a pixel scale of no larger than 3 μm. This ensures the morphological details and single cumulus cells are resolved as well as at least 2 pixels are available for the layer of cilia. A sensitivity of over 90 dB is generally preferred for the depth range that covers the oviduct tissue. A swept source OCT scheme is superior in minimizing the depthwise sensitivity drop. The B-scan imaging speed determines the highest frequency that can be detected for the CBF imaging, which is half of the sampling rate. For example, a B-scan rate of 100 Hz provides a highest measurable frequency of 50 Hz. It is known that the CBF in the mouse oviduct is generally smaller than 25 Hz. Thus, an imaging speed of at least 50 Hz is recommended for B-scans. Assuming

a 2 μm per pixel of the spatial sampling interval and a 1 mm scanning distance, 500 pixels are needed for the B-scan and therefore an A-line rate of at least 25 kHz is required for the OCT system. The capability to achieve higher imaging speed is preferred, since the CBF may exceed 25 Hz in studies with altered oviduct tissues. For building an OCT system, these specifications set primary requirements for the selections of the laser and the camera (or photodetector). The commercial OCT systems that meet such specifications are widely available in the market.

2. A dissection microscope with the magnification of around 4–5× is very useful for this step. A picture taken by the dissection microscope can also be used as a reference for the later analysis of the overall morphology of the oviduct.

3. A successful imaging of the oviduct 3D structure and CBF requires the targeted oviduct tissue to be stable. Tissue movement resulted from the breathing of the mouse can significantly reduce the imaging quality. Thus, the surgical clamps need to be very stably fixed to the lab jack, which will greatly help to immobilize the oviduct tissue.

4. To avoid strong surface reflection that could reduce the imaging quality, it is recommended to adjust the focal plane to be inside the oviduct tissue instead of on the surface of the tissue.

5. When setting the transverse scanning distance, always keep in mind the requirement of the pixel scale for the imaging of the oviduct, as suggested in **Note 1**.

6. Take average of B-scans from the same location can improve the quality (reduction of speckle noise) for 3D structural imaging of the oviduct.

7. When exporting the image files, it is recommended to set the dynamic range the same or larger than the range of the OCT signal amplitudes, thus to avoid saturation that will prevent certain quantitative analysis.

8. For imaging of the cumulus–oocyte complex inside the oviduct, timing could be critical. Monitoring of the mouse estrous cycle [16] is a good way to determine the time for performing the imaging experiments.

9. For the imaging result shown in Fig. 3, we utilized a sampling rate of 100 Hz and duration of 10.24 s for the B-scan imaging. The suggested B-scan repetition frequency can be found in **Note 1**. The duration of the B-scan imaging determines the spectral resolution, which is also the lowest frequency that can be resolved (~0.1 Hz in our case). Longer duration could be beneficial, however, with larger volumes of data to store and process, which will reduce the efficiency of imaging and can

increase the chance of being affected by the tissue motion. To ensure that the possible low CBFs can be covered, we recommend ~5–10 s for the duration of B-scan imaging.

10. Only the amplitude spectra from the positive frequency side will be utilized (the zero-frequency component should also be excluded from analysis).
11. The threshold of amplitude is set based on the noise level in the spectral domain. Since the mouse oviduct CBF is generally below 25 Hz, the parts of the amplitude spectra within the frequency range of 25–30 Hz are considered as noise and can be used to determine the amplitude threshold. Peaks with the amplitudes that are above the threshold are considered as the spectral information from the beat of cilia. For altered oviduct tissues where higher CBFs are expected, the frequency window where signals are taken as the noise should be shifted to a higher range accordingly. Also, the size of the window can be adjusted when necessary.
12. The thresholds of frequency are set as the lowest and the highest frequency positions of the peaks that represent the periodic ciliary activity.

Acknowledgments

This work was supported by the National Institute of Health grant R01HL120140 (I.V.L.) and the American Heart Association grant 16POST30990070 (S.W.).

References

1. Coy P, García-Vázquez FA, Visconti PE, Avilés M (2012) Roles of the oviduct in mammalian fertilization. Reproduction 144(6):649–660. https://doi.org/10.1530/rep-12-0279
2. Li S, Winuthayanon W (2017) Oviduct: roles in fertilization and early embryo development. J Endocrinol 232(1):R1–R26. https://doi.org/10.1530/joe-16-0302
3. Besenfelder U, Havlicek V, Brem G (2012) Role of the oviduct in early embryo development. Reprod Domest Anim 47(Suppl 4):156–163. https://doi.org/10.1111/j.1439-0531.2012.02070.x
4. Lyons RA, Saridogan E, Djahanbakhch O (2006) The reproductive significance of human fallopian tube cilia. Hum Reprod Update 12(4):363–372. https://doi.org/10.1093/humupd/dml012
5. Abe H, Oikawa T (1993) Observations by scanning electron microscopy of oviductal epithelial cells from cows at follicular and luteal phases. Anat Rec 235(3):399–410. https://doi.org/10.1002/ar.1092350309
6. Teilmann SC, Byskov AG, Pedersen PA, Wheatley DN, Pazour GJ, Christensen ST (2005) Localization of transient receptor potential ion channels in primary and motile cilia of the female murine reproductive organs. Mol Reprod Dev 71(4):444–452. https://doi.org/10.1002/mrd.20312
7. Bylander A, Nutu M, Wellander R, Goksör M, Billig H, Larsson DJ (2010) Rapid effects of progesterone on ciliary beat frequency in the mouse fallopian tube. Reprod Biol Endocrinol 8(1):1–8. https://doi.org/10.1186/1477-7827-8-48
8. Burton JC, Wang S, Stewart CA, Behringer RR, Larina IV (2015) High-resolution three-dimensional in vivo imaging of mouse oviduct using optical coherence tomography. Biomed

Opt Express 6(7):2713–2723. https://doi.org/10.1364/BOE.6.002713

9. Wang S, Burton JC, Behringer RR, Larina IV (2015) In vivo micro-scale tomography of ciliary behavior in the mammalian oviduct. Sci Rep 5:13216. https://doi.org/10.1038/srep13216. http://www.nature.com/articles/srep13216#supplementary-information
10. Fercher AF, Drexler W, Hitzenberger CK, Lasser T (2003) Optical coherence tomography—principles and applications. Rep Prog Phys 66(2):239
11. Schmitt JM, Xiang SH, Yung KM (1999) Speckle in optical coherence tomography. J Biomed Opt 4(1):95–105. https://doi.org/10.1117/1.429925
12. Cho A, Haruyama N, Kulkarni AB (2009) Generation of transgenic mice. In: Juan S. Bonifacino et al (eds) Current protocols in cell biology. Chapter19:Unit 19.11. John Willey & Sons, Inc., New Jersey. doi:https://doi.org/10.1002/0471143030.cb1911s42
13. Ploutarchou P, Melo P, Day AJ, Milner CM, Williams SA (2015) Molecular analysis of the cumulus matrix: insights from mice with O-glycan-deficient oocytes. Reproduction 149(5):533–543. https://doi.org/10.1530/REP-14-0503
14. Oldenburg AL, Chhetri RK, Hill DB, Button B (2012) Monitoring airway mucus flow and ciliary activity with optical coherence tomography. Biomed Opt Express 3(9):1978–1992. https://doi.org/10.1364/boe.3.001978
15. Wang S, Burton JC, Behringer RR, Larina IV (2016) Functional optical coherence tomography for high-resolution mapping of cilia beat frequency in the mouse oviduct in vivo. Proc SPIE 9689:96893R-96895
16. Caligioni C (2009) Assessing reproductive status/stages in mice. In: Jacqueline N. Crawley et al (eds) Current protocols in neuroscience. Appendix:Appendix-4I. John Willey & Sons, Inc., New Jersey. doi:https://doi.org/10.1002/0471142301.nsa04is48

Check for updates

Chapter 6

Live Imaging of Fetal Intra-abdominal Organs Using Two-Photon Laser-Scanning Microscopy

Yuhki Koike, Bo Li, Yong Chen, Hiromu Miyake, Carol Lee, Lijun Chi, Richard Wu, Mikihiro Inoue, Keiichi Uchida, Masato Kusunoki, Paul Delgado-Olguin, and Agostino Pierro

Abstract

The processes by which the intra-abdominal organ circulatory system develops in the embryo and during organogenesis are unclear. Previous studies have used fixed tissues to study the development of abdominal organ vasculature in the embryo; however, the intravital circulation of intra-abdominal organs in rodent fetal development has not been studied. This protocol describes a system that uses two-photon laser-scanning microscopy (TPLSM) for real-time observation and quantification of normal and pathologic live fetal intra-abdominal dynamics while the fetus is still connected to the mother via the umbilical cord.

Key words Two-photon laser scanning microscopy (TPLSM), Intravital imaging, Fetus

1 Introduction

Recent technical advances in in vivo microscopy have enabled us to visualize physiological and abnormal blood circulation in abdominal organs such as the liver, intestine and kidney in the adult mouse. However, visualization blood circulation in fetal pathology has not been possible mainly due to technical difficulty [1–5]. Zenclussen et al. recently published a technique to determine the developmental stage in mice using TPLSM to examine physiology of the uterus and placenta in vivo [6, 7]. The technique using TPLSM reported here does not require additional specialized equipment. Application of this intravital imaging system not only allows direct visualization of fetal organ development, but also to examine fetal intravital circulation. Additionally, this technique enables us to observe and quantify several parameters of the normal and pathologic physiology. This technology will be instrumental to further our understanding of pathologic processes that originate during embryogenesis.

Paul Delgado-Olguin (ed.), *Mouse Embryogenesis: Methods and Protocols*, Methods in Molecular Biology, vol. 1752, https://doi.org/10.1007/978-1-4939-7714-7_6,

2 Materials

This novel technique can be applied after embryonic day (E) 12.

2.1 Animals and Reagents

Experimental animals: GFP (constitutively expresses GFP) or *Rosa26*$^{mT/mG/+}$;*Tie2-Cre* (specifically expresses GFP in endothelial and hematopoietic cells) transgenic pregnant mouse at embryonic days 15–21 (E15–21). Embryos earlier than E14 cannot be imaged using this method in our hands due to technical difficulty.

- Isoflurane: 100% solution for inhalation anesthesia.
- Phosphate- buffered saline (PBS).
- Ultrapure water (e.g., Milli-Q, Millipore).
- Disinfection solution: 70% ethanol (volume/volume) in water.

2.2 Equipment

- Surgical instruments: dissecting scissors, blunt forceps, etc. (To reduce the risk of fetal injury, sharp forceps should not be used in this protocol.)
- Appropriate microscope stage with heating pad.
- Soft absorption pad.
- Syringes (1 ml) with more than 26-gauge needle.
- Gas anesthesia vaporizer.
- Oxygen gas.
- Anesthesia breathing circuit and nose cone.
- Kimwipes.
- Hair removal cream.
- Falcon tube (15 ml).
- Solder lug terminal; 0.3 mm; M4 (Manufacturer OSTERRATH).
- Cover glass.
- Holding devices of both cover glass and solder lug terminal: (TEKTON 7521 Helping Hand with Magnifier).

2.3 Microscope

- An inverted two-photon laser scanning microscope (TPLSM, e.g., Zeiss LSM710).
- Laser; Ti:Sapphire Chameleon Vision (Coherent).
- Objectives; 20× (Water immersion, Zeiss).
- Software application; ZEN (Zeiss, imaging software).

3 Methods

3.1 Surgical Setup for Intravital Imaging of Abdominal Cavity

1. The pregnant mouse is placed in the chamber for anesthetic induction (isoflurane mixed with 2% oxygen). Absence of the withdrawal reflex is confirmed with a toe pinch and examination of physiological responses (reduced respiration rate or heart beat).
2. The mouse is then transferred to the microscope stage, which has a heating pad.
3. To ensure the animal is unconscious during the procedure, isoflurane is continuously delivered through a cone covering the nose.
4. Abdominal hair is removed with hair removal cream.
5. The abdominal area is disinfected with 70% ethanol.
6. The Pfannenstiel incision (approximately 10–15 mm) is done in the lower abdominal area, and the uterus containing the fetus is gently externalized from the abdominal cavity using blunt forceps. A sterile gauze with a 10 mm diameter hole is put in the center to fix the uterus in position (Fig. 1a).

3.2 Fetus Setup for Positioning on the Microscope Stage

1. The fetus observation stage is prepared with disposable base molds (Fig. 1b). The edge of the base mold is cut following the orange dotted line (Fig. 1c). One of the edges is folded like the image in Fig. 1d. This folded area will be followed by the umbilical cord (blue arrow in Fig. 1d). The gauze is placed on the stage to provide a cushion for the fetus (red arrow in the Fig. 1e).
2. After the fetus observation stage is fixed on the mother's abdominal wall, the uterus is cut and the placenta is located (Fig. 1e, blue arrow). After making sure that the umbilical cord is straight so that blood flow is not affected (Fig. 1e, yellow arrow), the fetus is placed on the fetus observation stage.

3.3 Microscope Settings

1. Intravital observation is performed using a Zeiss LPM710 inverted microscope (Zeiss) with a 20× water immersion objective lens (W Plan-Apochromat 20×/1.0 DIC, VIS-IR M27 75 mm). The laser power is set up between 50 and 80%, and can be modified depending on the fetal age. Excitation wavelength used is 910 nm and scan speed is set to 12.7 μs/pixel.
2. The fixing device was made using a solder lug terminal, cover glass, and a TEKTON 7521 Helping Hand (Fig. 2a).
3. The fixing device is gently placed on top of the abdominal wall of the fetus.
4. After confirming that the fixing device is stable, one drop of water is placed onto the cover glass (Fig. 2b), which has to be attached to the objective lens (Fig. 2c).

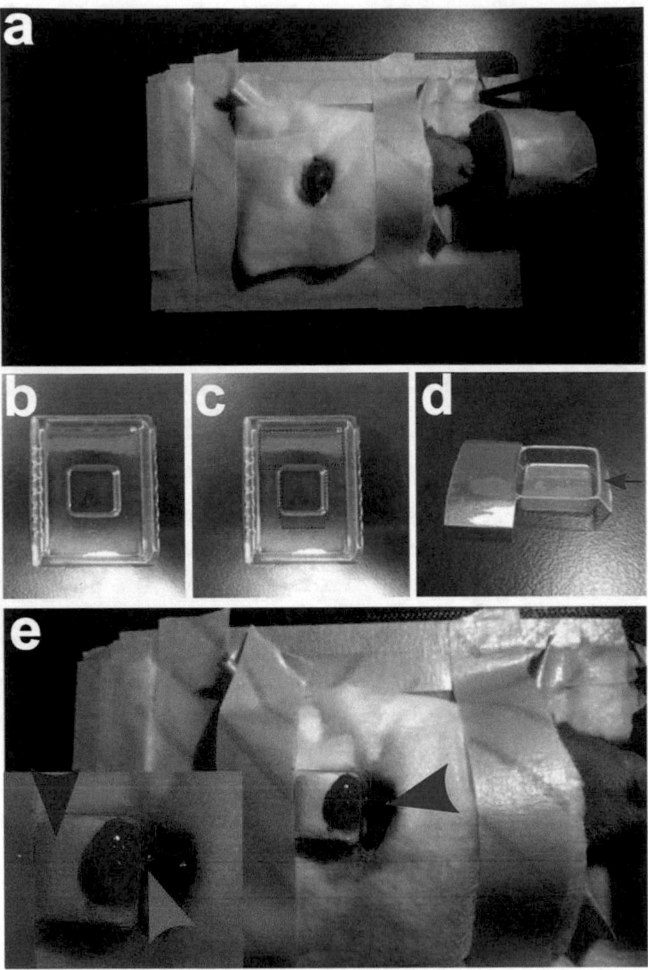

Fig. 1 Mouse and fetus fixation on the specific hand-made stage of the two-photon laser scanning microscope. (**a**) The pregnant mouse is fixed on the stage with a heating pad, and after exteriorizing the uterus containing one fetus out of the abdominal cavity, a sterile gauze with a 10 mm hole in the center is used to fix the position of the uterus. (**b–d**) The special hand-made fetal platform is made of a normal plastic base mold case. The blue arrow indicates the direction of the umbilical cord from the placenta when the fetus is put on the stage. (**e**) Image of the set up. The blue arrow is pointing to the placenta, the yellow arrow indicates the umbilical cord, and the red arrow is the soft cushion that is put on the hand-made fetal platform

5. Observation by two-photon excitation (Fig. 2d) is conducted, and the acquisition data is saved. Saving as a czi file type is recommended because this file type can be used for the addition of a scale bar or to create a movie.
6. The pregnant female and fetuses are then euthanized following guidelines from the corresponding Animal Care Committee.

3.4 Data Processing

1. The ZEN lite software (Zeiss Corporation) is launched, the saved czi files are opened, and the scale bar is added. To obtain data from sequential time course images, choose the image to

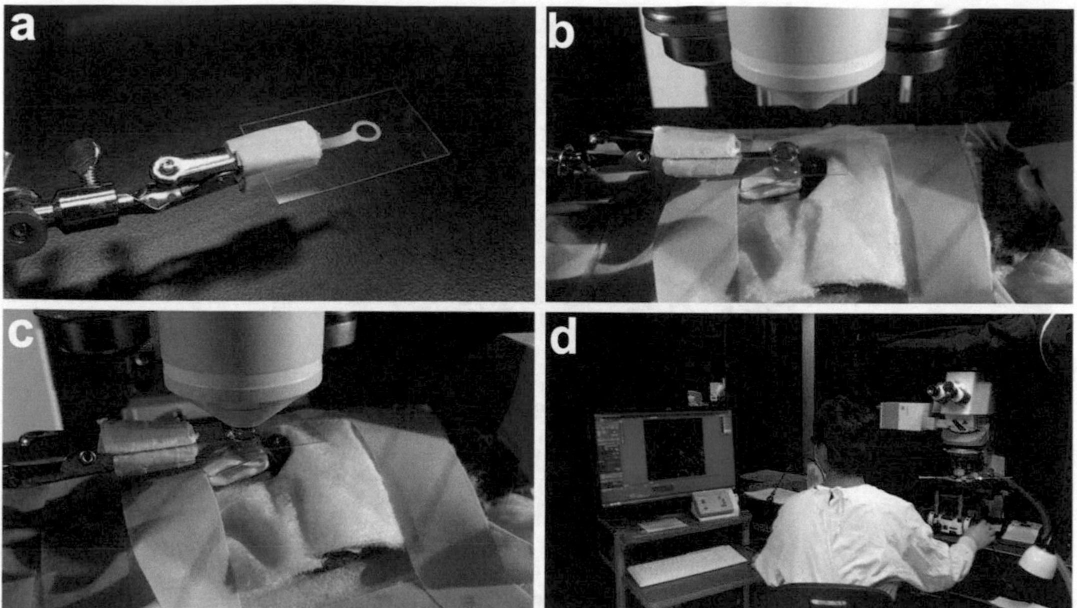

Fig. 2 Microscope set up. (**a**) The fetal fixation device is made from a solder lug terminal, cover glass, and commercially available TEKTON 7521 Helping Hand. (**b**) Observation is performed using a water immersed 20× objective lens. (**c**, **d**) Image of the microscope and the environment in which this protocol takes place

be analyzed and select "Export/Import" from the file tag, select "export," then select the appropriate file type (JPEG, TIFF, PNG, etc.), and then "Apply." To make a movie, select "movie export" and choose the appropriate file type (AVI, WMF, MOV, etc.) (Fig. 3).

4 Notes

1. Uncouple the normal sized anesthetic cone and place one finger of a rubber glove over the edge of the anesthetic breathing circuit. Cut a hole at the edge of the rubber finger to fit the mouth of the pregnant mouse. It is important for the anesthetic cone to fit snugly around the mouse to reduce isoflurane gas leakage and perform general anesthesia properly.
2. Preparations prior to preparation of specimens, including turning on the heating pad before starting the procedure, are important to reduce the time under anesthesia for the animal and stress.
3. Confirming the absence of the withdrawal reflex as well as physiological responses should be performed at least every 10 min during the procedure to ensure deep anesthesia.
4. After removing the abdominal hair, cleaning the residual removed hair on the abdominal skin is important to prevent hair from contaminating the observation area.

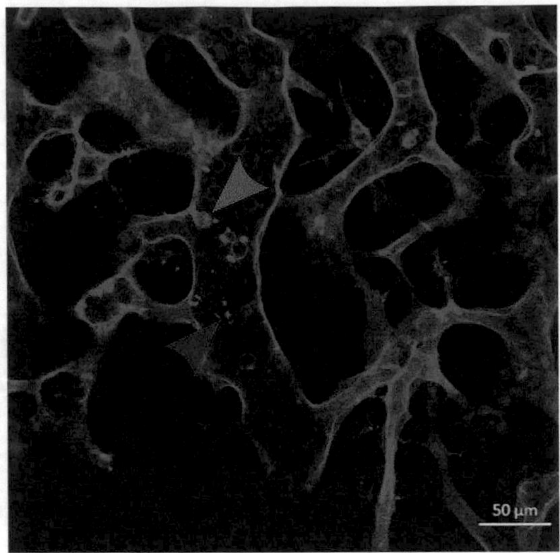

Fig. 3 An E14 image of the intra-abdominal cavity in the live fetus while keeping continuous blood supply from the umbilical cord. The yellow arrow indicates leukocytes on the blood vessel wall. The red arrow indicates platelets. Fetal red blood cells were seen as round-shaped components with the faded color in the center. The scale bar is 50 μm in length

5. A sterilized bordered gauze should be used on the disinfected abdominal area to keep the operation area clean (Fig. 1a). A nonsterilized operation procedure may modify the results due to infection.
6. Gentle maneuver is always necessary when transferring the uterus out of the abdominal cavity. Using forceps, grasp the selected segment of the uterus with as many contact points as possible between the tissue and forceps, to avoid puncturing or injuring the uterus or fetus.
7. When preparing the fetus observation stage, the fold made on one of the edges (blue arrow in Fig. 1d) should be smooth. A sharp folded edge may block blood circulation through the umbilical cord. When placing the fetus observation stage on the maternal abdomen, the bottom of the stage should be completely in contact with the maternal abdominal wall through a sterilized gauze to keep the fetal temperature warm.
8. It is *very* important to never damage the umbilical cord, otherwise fetal circulation will be lost. The distance between the fetus and placenta should be minimized to reduce tension in the umbilical cord and maintain fetal circulation.
9. A very high laser power will burn the fetal skin at the observation area. Suitable laser power will depend on fetal age. A higher laser power will be needed for more advanced gestational ages due to increased abdominal wall depth.

10. The size of the solder lug terminal can be selected per the fetus body size.
11. When putting the fixing device on the fetal abdominal wall, limit formation of air pockets between the abdominal wall and the cover glass as much as possible. If there is an air pocket, put one drop of warmed distilled water on the fetal abdominal wall to reduce its size.
12. Too much compression on the fetal abdominal wall by the fixing device may cause insufficient fetal intra-abdominal circulation and also induce intra-abdominal visceral injury. Gentle compression is essential.
13. A quick scan and recording is necessary to reduce the stress of both the pregnant mice and the fetus. A longer observation time will decrease fetal circulation due to drop of body temperature.
14. If the first observed fetal image quality is not acceptable, you may want to move on to the next uterus. Circulating blood cells should be clearly observed (Fig. 2). Both the uterus and fetus in the abdominal cavity usually have well-conditioned circulation and temperature during the first fetal observation period, however, the whole procedure should be completed within 2 h to reduce the effects of anesthesia on the fetus and pregnant mouse circulation.
15. The ZEN lite software can be downloaded free from the Zeiss microscope website after registering.
16. Using the saved czi files, you can generate three dimensional intra-abdominal fetal images from the z-stacked sequential images.

References

1. Koike Y, Tanaka K, Kobayashi M, Toiyama Y, Inoue Y, Mohri Y et al (2015) Dynamic pathology for leukocyte-platelet formation in sepsis model. J Surg Res 195(1):188–195
2. Tanaka K, Koike Y, Shimura T, Okigami M, Ide S, Toiyama Y et al (2014) In vivo characterization of neutrophil extracellular traps in various organs of a murine sepsis model. PLoS One 9(11):e111888
3. Mizuno R, Kamioka Y, Kabashima K, Imajo M, Sumiyama K, Nakasho E et al (2014) In vivo imaging reveals PKA regulation of ERK activity during neutrophil recruitment to inflamed intestines. J Exp Med 211(6):1123–1136
4. Koike Y, Uchida K, Tanaka K, Ide S, Otake K, Okita Y et al (2014) Dynamic pathology for circulating free DNA in a dextran sodium sulfate colitis mouse model. Pediatr Surg Int 30(12):1199–1206
5. Dunn KW, Sandoval RM, Molitoris BA (2003) Intravital imaging of the kidney using multiparameter multiphoton microscopy. Nephron Exp Nephrol 94(1):e7–11
6. Zenclussen AC, Olivieri DN, Dustin ML, Tadokoro CE (2013) In vivo multiphoton microscopy technique to reveal the physiology of the mouse uterus. Am J Reprod Immunol 69(3):281–289
7. Zenclussen AC, Olivieri DN, Dustin ML, Tadokoro CE (2012) In vivo multiphoton microscopy technique to reveal the physiology of the mouse placenta. Am J Reprod Immunol 68(3):271–278

Check for updates

Chapter 7

Embryonary Mouse Cardiac Fibroblast Isolation

Alejandra Garate-Carrillo and Israel Ramirez

Abstract

Mouse cardiac fibroblasts have been widely used as an in vitro model for studying fundamental biological processes and mechanisms underlying cardiac pathologies, as well as identifying potential therapeutic targets. Cardiac FBs are relatively easy to culture in a dish and can be manipulated using molecular and pharmacological tools. Because FBs rapidly decrease cell cycle division and proliferative rate after birth, they are prone to phenotypic changes and senescence in cell culture soon after a few passages. Therefore, primary cultures of differentiated fibroblasts from embryos are more desirable. Below we will describe a method that provides good cell yield and viability of E16 CD-1 mouse embryonic cardiac fibroblasts in primary cultures.

Key words Cardiac fibroblasts, Heart, Enzymatic digestion, Embryonic mice, Isolation, Cell culture

1 Introduction

Seminal studies demonstrated that the mouse heart consists of approximately 30% cardiac myocytes and 70% nonmyocytes. Being fibroblast the majority of these cells [1–6].

The definition of the fibroblast is based on morphological characteristics (elongated fiber shape) that can vary depending on the location within the organism. Morphologically, fibroblasts are flat, spindle-shaped cells which lack a basement membrane [1] and feature a round, elongated nucleus surrounded by vast endoplasmic reticulum [7]. There are different theories regarding the origin of the cardiac fibroblast lineage. In the mammalian embryo, cardiac fibroblasts are thought to derive primarily from the proepicardial organ via epithelial-to-mesenchymal transition (EMT) [7–14], although there might be some subsequent contribution from bone marrow-derived precursors [1, 7, 15]. In the adult, it is not well established how the cardiac fibroblast population is maintained during normal homeostasis, but it is thought that resident

Paul Delgado-Olguin (ed.), *Mouse Embryogenesis: Methods and Protocols*, Methods in Molecular Biology, vol. 1752, https://doi.org/10.1007/978-1-4939-7714-7_7, © Springer Science+Business Media, LLC, part of Springer Nature 2018

fibroblasts and epithelial cells undergoing EMT are the primary source of the low turnover within the adult heart [1, 7, 8, 16]. To date, fibroblasts have been difficult to accurately identify. In some cases, fibroblasts are identified based on their spindle shape combined with positive staining for the mesenchymal marker vimentin and the absence of staining for markers of epithelial or other cell types, such as smooth muscle cells, astrocytes, or hematopoietic cells [17, 18].

Cardiac fibroblasts have been termed "sentinel" cells, as they can sense changes in chemical, mechanical and electrical signals in the heart, and trigger an appropriate response. One of the primary functions of the cardiac fibroblast is the regulation of synthesis and degradation of the extracellular matrix (ECM) in the myocardium to provide a three-dimensional network for myocytes and other cells of the heart. In addition, cardiac fibroblasts are involved in synthesis and secretion of growth factors and cytokines [1]. In fact, under physiological condition cardiac fibroblasts not only produce and remodel the ECM in response to different physiological cues [7, 8, 19, 20], but are also involved in several tissue functions such as regulation of growth during development and proliferation of the cardiomyocyte [21, 22]. Fibroblasts directly connect to cardiomyocytes via connexins [4, 7, 23, 24], and are able to electrically isolate various portions of the conduction system in the heart [8, 18, 25] and secrete factors that signal to cardiomyocytes in a paracrine fashion [1, 7, 8, 19, 20, 26]. Meanwhile under pathological conditions fibroblasts can induce cardiomyocyte hypertrophic and fibrotic responses to injury in the adult heart [7, 18, 20] by secreting cytokines (TNFα, pro-inflammatory interleukins and TGFβ), active peptides (angiotensin II, endothelin 1), and growth factors [7, 27], which stimulate the myocardium in autocrine and/or paracrine fashions [28].

Hence, cardiac fibroblasts play an important role in normal and pathologic performance of the heart. An understanding of the fibroblast biology requires in depth studies using a stable and reliable system in which the biological responses to different stimuli can be determined [29]. Thus, the deeper our understanding of fibroblast behavior within a given environment, the more capable we will be to modulate their responses [7]. Given their several roles within the heart, cardiac fibroblasts provide an excellent model system to study many aspects of heart function and pathophysiology, including the regulation of cell growth, survival, inflammation, hormone secretion and metabolism [30]. To unravel the molecular and cellular mechanisms for normal cardiac function and cardiac pathology, high-quality isolated cardiac fibroblasts are essential [31]. Although there are several methods to obtain fibroblast of different animal species there is no consensus on the optimal protocol to isolate and culture viable cells [31]. However, at their core, the available protocols generally rely on digestion of the heart with an enzymatic solution that will disrupt the hardest tissue releasing into the media the cellular component of the heart [31].

Nevertheless, the culture of isolated fibroblasts is an important yet difficult point of the process, because as primary cells, they cannot be cultured indefinitely due to onset of replicative senescence or aneuploidization, leaving just a few passages and little time to set up experiments and get results. Therefore, new techniques or improved protocols are needed to extend the life span of the cultures [32].

Isolation of embryonic fibroblasts now produces better yields and longer times of viability. Jonsson et al., defined the transcriptomic, morphologic and epigenetic differences between fetal and adult human cardiac fibroblast (fHCFs and aHCFs) using cellular, molecular, and genome-wide sequencing approaches [33]. The results showed that fHCFs are smaller and proliferate more quickly than their adult counterpart and express genes that reveal their role in heart growth. In contrast, the aHCF transcriptome revealed a function in maintenance of the mature heart [33]. Such differences between adult and fetal fibroblasts might influence their viability and suggest that embryonic cardiac fibroblast could be more desirable for molecular and pathophysiology studies of the heart.

Embryonic fibroblasts have been used as feeder cells to maintain stem cells in culture because fibrobasts secrete ECM, which helps keeping the stem cells in an undifferentiated state [34]. Another reason to isolate embryonic fibroblasts is that they can be transdifferentiated into a variety of mature cells such as cardiomyocytes [35], providing different models to study development or cardiac diseases.

This protocol is useful for embryonic mouse cardiac fibroblast isolation, and its primary aim is to maintain a sufficient numbers of viable of cells to study cell signaling pathways that could potentially lead to cardiac fibrosis [1].

2 Materials

Prepare all solutions using ultrapure water (prepared by purifying deionized water) and analytical grade reagents. Prepare and store all reagents at room temperature, unless indicated otherwise. Follow laboratory animal care and biological waste disposal regulations at all time.

2.1 ADS Buffer Preparation

1. 1 L enzyme-free ADS buffer stock solution: Add about 300 mL ultrapure water to a 1 L glass beaker. Weight enough amount to prepare 1 L of 20 mM HEPES (free acid) and transfer to beaker to start dissolving, 0.8 mM $MgSO_4$ (anhydrous), 1 mM NaH_2PO_4, 116 mM NaCl, 180.16 mM D-glucose, 5.4 mM KCl, 0.02 g phenol red indicator, and 1% of penicillin–streptomycin–fungizone (PSF) antibiotic–antimycotic mixture solution (*see* **Note 1**).
2. Adjust pH to 7.35 with NaOH. Set the volume to 1000 mL with ddH_2O and mix to dissolve all reagents.
3. Filter buffer solution inside the hood to avoid contamination and store in a labeled sterile bottle at 4 °C (*see* **Note 2**).
4. Enzyme mixture must be prepared fresh and filtered.
   (a) Ads buffer for 100 mL.
   (b) Collagenase 0.625% (w/v) 0.0625 g (~130 U/mL).
   (c) Pancreatin 0.6% (w/v) 0.060 g.
5. Place the mixture on the stirring plate for 3 h to dissolve the enzymes. Once the enzyme mixture is completely dissolved, sterilize by filtration and then the buffer is ready to use. Enzyme activity may vary depending on the source. If you experience trouble with digestion, enzyme mixture should be titered.

2.2 Equipment

Surgical kit (autoclaved).

50–100 mL spinner flasks (autoclaved).

37 °C water bath.

Hot stirring plate.

2× 10 cm plates with enzyme-free ADS buffer.

1× 10 cm plates.

Fetal bovine serum (FBS).

Dulbecco's Modified Eagle's Media (DMEM) complete media (10% FBS + 1% PSF).

3 Methods

3.1 Procedure

1. CD-1 mice need to be time-mated using vaginal plug to designate embryonic stage. At day 20 sedate pregnant mouse with isoflurane (0.5%) using a gas chamber. Shave the abdominal region and clean it with 70% ethanol, open it with scissors and collect the whole litter (8–10 embryos).
2. Proceed to open the embryo chest using scissors, localize hearts and excise tissue, then remove aortas and atrial tissue and cut the heart in small pieces. Place cardiac tissue into a

Fig. 1 Embryonic hearts being enzymatically digested. Hearts are placed in a spinning flask with 10 mL of ADS buffer incubated and stir in water bath at 37 °C

10 cm cell culture dish with enzyme-free ADS media to eliminate blood excess and clots. Transfer to 6 cm plate with enzyme-free ADS buffer.

3. Transfer cardiac tissue portions into a spinning 50–100 mL flask with 10 mL of fresh active ADS buffer (Type II collagenase and pancreatin) over a stir plate with water bath at 37 °C for 5 min to digest tissue. Using a 2 mL pipette, carefully discard the supernatant.
4. Add 5 mL of fresh active ADS buffer to spinning flask and incubate in water bath for 10 min at 37 °C (Fig. 1). After digestion, transfer all the recovered supernatant into a 15 mL falcon tube with 1 mL of FBS to stabilize cells (*see* **Note 3**). Centrifuge for 5 min at 1500 × *g*, discard the supernatant and resuspend pellet in 1 mL FBS; place it into the incubator at 37 °C (*see* **Note 4**).
5. Repeat the last step five times or until tissue becomes watery. Each 10 min digestion, recover the pellet, and resuspend it in 1 mL FBS. After five incubations, pool all the resulting pellets in a 50 mL falcon tube.

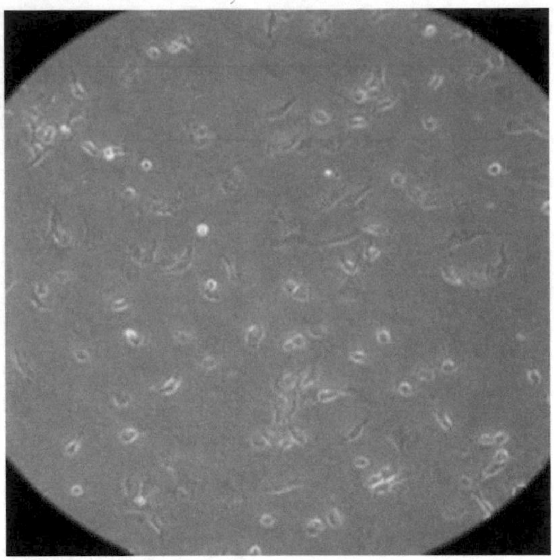

Fig. 2 Microphotograph of isolated cardiac embryonic fibroblast in culture after seven days of isolaton

3.2 Plating Cells

1. Centrifuge the pooled cells for 5 min at 1500 × *g*, discard supernatant.
2. Resuspend pool pellet by adding 7 mL of DMEM complete media. Centrifuge at 1500 × *g* for 5 min.
3. Resuspend pellet with complete DMEM media and plate onto 1× 10 cm dish. Incubate for 1 h at 37 °C in 5% CO_2.
4. After 1 h of incubation, get dish out of incubator, swirl the dish clockwise, replace the media, and then incubate 37 °C in 5% CO_2.

Next day shake the plate gently to strip myocytes off; at this time FBs must be attached to the dish forming a layer. Replace the media every day, wait until cells reach confluency, and maintain cell culture by standard conditions (Fig. 2).

3.3 Passage and Freezing Cells

1. When confluent split the cells dividing one 10 cm dish into 3 10 cm plates (*see* **Note 5**).
2. To freeze cells divide 1 10 cm dish onto two vials.
3. Few days after the extraction fibroblasts can be identified by immunofluorescence for vimentin (Fig. 3).

4 Notes

1. These quantities are good enough for a few experiments, as correspond to more than 100 embryonic hearts. (From one pregnant mouse eight to ten embryonic hearts can be extracted.)

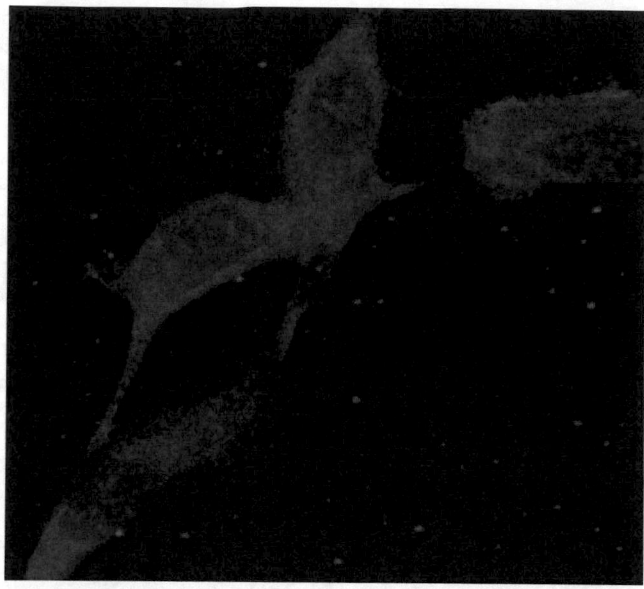

Fig. 3 Vimentin immunofluorescence for identification of fibroblasts. Spindle-like shape, with rounded nucleus is one of their main morphologic characteristics

2. Ads buffer without adding the enzymes could be prepared in advance and be store for several days.
3. When collecting digestion supernatant use the hood to avoid contamination and avoid aspirating the tissue at all times.
4. Digestion must be at 37 °C in order to maintain enzyme activity.
5. When spliting cells, if they are not growing very fast, they can be seeded into a 6 cm dish or 6-well plates depending on the experiment.

References

1. Souders C, Bowers S, Baudino TA (2012) Cardiac fibroblast: the renaissance. Cell 105(12):1164–1176
2. Nag AC (1980) Study of non-muscle cells of the adult mammalian heart: a fine structural analysis and distribution. Cytobios 28(109):41–61
3. Zak R (1974) Development and proliferative capacity of cardiac muscle cells. Circ Res 32(Suppl 2):17–26
4. Baudino TA, Carver W, Giles W, Borg TK (2006) Cardiac fibroblasts: friend or foe? Am J Physiol Heart Circ Physiol 291:H1015–H1026
5. Camelliti P, Borg TK, Kohl P (2005) Structural and functional characterization of cardiac fibroblasts. Cardiovasc Res 65:40–51
6. Harvey PR, Rosenthal N (1999) Heart development, vol 1. Academic, New York
7. Lajiness JD, Conway SJ (2012) The dynamic role of cardiac fibroblasts in development and disease. J Cardiovasc Translat Res 5(6):739–748
8. Snider P, Standley KN, Wang J, Azhar M, Doetschman T, Conway SJ (2009) Origin of cardiac fibroblasts and the role of periostin. Circ Res 105(10):934–947
9. Gittenberger-de Groot AC, VPM MMMT, Gourdie RG, Poelmann RE (1998) Epicardium-derived cells contribute a novel population to the myocardial wall and the atrioventricular cushions. Circ Res 82(10):1043–1052

10. Kolditz DP, Wijffels MC, Blom NA, van der Laarse A, Hahurij ND, Lie-Venema H, Markwald RR, Poelmann RE, Schalij MJ, Gittenberger-de Groot AC (2008) Epicardium-derived cells in development of annulus fibrosis and persistence of accessory pathways. Circulation 117(12):1508–1517
11. Mikawa T, Gourdie RG (1996) Pericardial mesoderm generates a population of coronary smooth muscle cells migrating into the heart along with ingrowth of the epicardial organ. Dev Biol 174(2):221–232
12. Perez-Pomares JM, Carmona R, Gonzalez-Iriarte M, Atencia G, Wessels A, Munoz-Chapuli R (2002) Origin of coronary endothelial cells from epicardial mesothelium in avian embryos. Int J Dev Biol 46(8):1005–1013
13. Lie-Venema H, van den Akker NM, Bax NA, Winter EM, Maas S, Kekarainen T, Hoeben RC, deRuiter MC, Poelmann RE, Gittenberger-de Groot AC (2007) Origin, fate, and function of epicardium-derived cells (EPDCs) in normal and abnormal cardiac development. Sci World J 7:1777–1798
14. Wessels A, van den Hoff MJ, Adamo RF, Phelps AL, Lockhart MM, Sauls K, Briggs LE, Norris RA, van Wijk B, Perez-Pomares JM, Dettman RW, Burch JB (2012) Epicardially derived fibroblasts preferentially contribute to the parietal leaflets of the atrioventricular valves in the murine heart. Dev Biol 366(2):111–124
15. Norris RA, Borg TK, Butcher JT, Baudino TA, Banerjee I, Markwald RR (2008) Neonatal and adult cardiovascular pathophysiological remodeling and repair: developmental role of periostin. Ann New York Acad Sci 1123:30–40
16. Krenning G, Zeisberg EM, Kalluri R (2010) The origin of fibroblasts and mechanism of cardiac fibrosis. J Cell Physiol 225(3):631–637
17. Chang HY, Chi JT, Dudoit S, Bondre C, van de Rijn M, Botstein D, Brown PO (2002) Diversity, topographic differentiation, and positional memory in human fibroblasts. Proc Natl Acad Sci U S A 99:12877–12882
18. Goodpaster T, Legesse-Miller A, Hameed MR, Aisner SC, Randolph-Habecker J, Coller HA (2008) An immunohistochemical method for identifying fibroblasts in formalin-fixed, paraffin-embedded tissue. J Histochem Cytochem 56(4):347–358
19. Snider P, Hinton RB, Moreno-Rodriguez RA, Wang J, Rogers R, Lindsley A, Li F, Ingram DA, Menick D, Field L, Firulli AB, Molkentin JD, Markwald R, Conway SJ (2008) Periostin is required for maturation and extracellular matrix stabilization of noncardiomyocyte lineages of the heart. Circ Res 102(7):752–760
20. Takeda N, Manabe I (2011) Cellular interplay between cardiomyocytes and nonmyocytes in cardiac remodeling. Int J Inflam 2011:535241
21. Ieda M, Tsuchihashi T, Ivey KN, Ross RS, Hong TT, Shaw RM, Srivastava D (2009) Cardiac fibroblasts regulate myocardial proliferation through beta1 integrin signaling. Dev Cell 16(2):233–244
22. Noseda M, Schneider MD (2009) Fibroblasts inform the heart: control of cardiomyocyte cycling and size by age-dependent paracrine signals. Dev Cell 16(2):161–162
23. Gaudesius G, Miragoli M, Thomas SP, Rohr S (2003) Coupling of cardiac electrical activity over extended distances by fibroblasts of cardiac origin. Circ Res 93(5):421–428
24. Miragoli M, Gaudesius G, Rohr S (2006) Electrotonic modulation of cardiac impulse conduction by myofibroblasts. Circ Res 98(6):801–810
25. Spach MS, Boineau JP (1997) Microfibrosis produces electrical load variations due to loss of side-to-side cell connections: a major mechanism of structural heart disease arrhythmias. Pacing Clin Electrophysiol 20(2 Pt 2): 397–413
26. Ottaviano FG, Yee KO (2011) Communication signals between cardiac fibroblasts and cardiac myocytes. J Cardiovasc Pharmacol 57(5):513–521
27. Wang D, Oparil S, Feng JA, Li P, Perry G, Chen LB, Dai M, John SW, Chen YF (2003) Effects of pressure overload on extracellular matrix expression in the heart of the atrial natriuretic peptide null mouse. Hypertension 42(1):88–95
28. Dong F, Abhijit T, Jiwon L, Zamaneh K (2012) Cardiac fibroblasts, fibrosis and extracellular matrix remodeling in heart disease. Fibrogen Tissue Rep 5(1):15
29. Agocha AE, Eghbali-Webb M (1997) A simple method for preparation of cultured cardiac fibroblasts from adult human ventricular tissue. Mol Cell Biochem 172(1–2):195–198
30. Cell applications (.1994-2017) Rat cardiac fibroblasts: RCF. https://www.cellapplications.com/rat-cardiac-fibroblasts-rcf. Accessed 11 Feb 2017
31. Chen X, O'Connell TD, Xiang YK (2016) With or without langendorff: a new method for adult myocyte isolation to be tested with time. Circ Res 119(8):888–890
32. Seluanov A, Vaidya A, Gorbunova V (2010) Establishing primary adult fibroblast cultures from rodents. J Vis Exp 5(44). https://doi.org/10.3791/2033. http://www.jove.com/details.php?id=2033

33. Jonsson MKB, Hartman RJG, Ackers-Johnson M, Tan WLW, Lim B, van Veen TAB, Foo RS (2016) A transcriptomic and epigenomic comparison of fetal and adult human cardiac fibroblasts reveals novel key transcription factors in adult cardiac fibroblasts. JACC Basic Transl Sci 1(7):590–602

34. Wu HY, Opperman K, Kaboord B (2014) An improve method for highly efficient isolation of primary mouse embryonic fibroblasts. .Thermo Fisher Scientific. https://www.thermofisher.com/mx/es/home/life-science/protein-biology/protein-biology-learning-center/protein-biology-resource-library/protein-biology-application-notes/method-efficient-isolation-primary-mouse-embryonic-fibroblasts.html. Accessed 22 Feb 2017

35. Efe JA, Hilcove S, Kim J, Zhou H, Ouyang K, Wang G, Chen J, Ding S (2011) Conversion of mouse fibroblasts into cardiomyocytes using a direct reprogramming strategy. Nat Cell Biol 13(3):215–222

Chapter 8

Explant Culture for Studying Lung Development

Behzad Yeganeh, Claudia Bilodeau, and Martin Post

Abstract

Lung development is a complex process that requires the input of various signaling pathways to coordinate the specification and differentiation of multiple cell types. Ex vivo culture of the lung is a very useful technique that represents an attractive model for investigating many different processes critical to lung development, function, and disease pathology. Ex vivo cultured lungs remain comparable to the in vivo lung both in structure and function, which makes them more suitable than cell cultures for physiological studies. Lung explant cultures offer several significant advantages for studies of morphogenetic events that guide lung development including budding, branching, and fusion. It also maintains the native physiological interactions between cells in the developing lung, enabling investigations of the direct and indirect signaling taking place between tissues and cells throughout the developmental process. Studying temporal and spatial control of gene expression by transcriptional factors using different reporters to understand their regulatory function at different moments of development is another valuable advantage of lung explants culture.

Key words Lung branching morphogenesis, Organ culture, Air–liquid interface, Lung slice, Ex vivo culture

1 Introduction

The idea of culturing tissue or an organ outside of the body, a technique called explant culture, has its origin at the beginning of the twentieth century. The first studies using ex vivo cultures have shown that embryological tissues differentiate and behave like what appears to be normal tissue, indicating that we can use this technique to investigate fetal and embryological development [1].

Explant culture allows the study of many different processes implicated in organogenesis. A major advantage of this technique is that it maintains the native physiological interactions between cells in the tissue, enabling observation of the direct and indirect signaling taking place between tissues and cells throughout the developmental process [2]. This approach also facilitates characterization of different morphogenetic events that lead to embryonic organ development including budding, branching and fusion. At a

Paul Delgado-Olguin (ed.), *Mouse Embryogenesis: Methods and Protocols*, Methods in Molecular Biology, vol. 1752, https://doi.org/10.1007/978-1-4939-7714-7_8,

cellular level it is possible to look more closely and target migration, proliferation and differentiation of specific cells [3]. Ex vivo culture of explants are also useful to follow gene expression identified by different reporters and understand their regulatory function as well as identifying signaling molecules secreting at different moments of development [3].

Due to their size, the entire organ of an embryo can be used as an explant. It is suggested that such explants are more resistant to hypoxic conditions than mature tissue. Therefore, the technique of ex vivo explants culture has predominantly been used for developmental studies [4]. However, in 1959, Trowel developed a protocol to culture a mature organ [1, 4]. This method is useful to study differences in pharmacological response in order to develop new therapies. It has also been used to observe the reaction of the tissue to infection [5].

Obviously there are some limitations to the explant technique. Unfortunately, an organ alone does not reproduce the physiological environment as a full body, and they are not suitable for long term experiments. Under these circumstances, organ transplantation is a more appropriate alternative. However, the ex vivo culturing technique offers great advantages: researchers are able to control the media, manipulate the tissue and the environment to create the perfect circumstances to observe the effect of various growth factors and molecules under reproducible conditions [3].

The explant technique has been applied to many organs, including the lung. Embryonic lung explants adopt a flattened morphology that allows quantification of terminal branches and branch length. This technique has allowed the study of some of the key regulators implicated in budding and branching of the lung like fibroblast growth factor (FGF), bone morphogenetic protein (BMP) and wingless type (Wnt) signaling family [6]. Also, lung embryogenesis is closely regulated by epithelial and mesenchymal interactions and ex vivo cultures have helped to better characterize growth factors and signaling molecules implicated in these interactions [6]. Ex vivo explants could be established using the entire lung or only the mesenchymal and epithelial components in order to further characterize their functions [6]. Explant culture has been and will continue to be instrumental to further our knowledge on the cellular and molecular mechanisms controlling lung development.

2 Materials

2.1 Embryonic (E11.5) Mouse Lung Explant Culture

1. Dulbecco's Modified Eagle's Medium: Nutrient Mixture F-12 (DMEM/F-12).
2. Hank's Buffered Salt Solution ($HBSS^-$, no calcium, no magnesium).

3. Penicillin–streptomycin.
4. L-ascorbic acid.
5. Nuclepore Track-Etched Polycarbonate Membrane Filter (8 μm pore size, Whatman)
6. Plastic disposable transfer pipets, Sterile.
7. Flat bottom cell culture plates with lids, 6-well plates.
8. Stereoscopic dissecting microscope.
9. Dissection tools including Fine iris scissors (straight), Noyes spring scissors and Dumont forceps.
10. 50 ml conical centrifuge tubes.
11. Fisherbrand™ petri dishes with clear lid, 100 mm × 15 mm.
12. Insulin syringe with detachable needle.

2.2 E17.5 and Adult Mouse Lung Explant Culture

1. Low melting point agarose.
2. Dulbecco's Modified Eagle's Medium: Nutrient Mixture F-12 (DMEM/F-12).
3. Hank's Buffered Salt Solution (HBSS).
4. Dextrose (D-glucose) anhydrous.
5. Gibco™ Penicillin/streptomycin.
6. L-ascorbic acid.
7. Nucleopore hydrophobic floating membranes (8 μm pore size, Whatman)
8. Dissection tools including Fine iris scissors (straight), Noyes spring scissors and Dumont forceps.
9. Flat bottom cell culture plates with lids, 6-well plates.
10. Stereoscopic dissecting microscope.
11. Vibrating blade microtome.
12. 50 ml conical centrifuge tubes.
13. Petri dishes with clear lid, 100 mm × 15 mm.

Animal experiments were carried out in accordance with the Animal Care Committee guidelines of the Hospital for Sick Children Research Institute.

3 Methods

Development of the lung in the mouse embryo is a complex process that begins at embryonic day 9.5 (E9.5), upon initiation of the respiratory diverticulum or lung buds from the ventral foregut endoderm and continues throughout the early postnatal stages. Based on histological morphology, mouse lung development and maturation has been divided into five stages: embryonic stage

(E9–12), pseudoglandular (E12–14.5), canalicular (E14.5–16.5), terminal saccular (E16.5 to PD5), and alveolar or postnatal stage (PD5 to PD30). In order to determine the key molecules involved in the regulation of lung branching morphogenesis, lung organ culture can be performed at different stages of lung development. In the following section, we will describe two protocols used for culturing lung explants at early (E11.5) and late (E17.5) lung developmental stages.

3.1 Embryonic Lung Isolation and Culture

1. Set up timed mating with embryonic day zero (E0) defined as the day of vaginal plug. On postcoitum day 11.5 (E11.5), euthanize the timed-pregnant mice using CO_2 (*see* **Note 1**)
2. Under sterile conditions with appropriate skin disinfectant (*see* **Note 2**) (the abdomen area could be disinfected with 70% ethanol), remove the uterus from the animal in a laminar flow hood and place it in a 50 ml falcon tube containing precooled Hanks' Balanced Salt Solution (HBSS) on ice.
3. Place the uterus in a Petri dish containing cold HBSS under the stereoscopic dissecting microscope and release the embryos from the uterus (*see* **Note 3**) (by incising the uterine wall using sterile spring scissors).
4. Transfer the embryos using a sterile squeezable Transfer Pipette in a new Petri dish containing cold HBSS and place it on ice (*see* **Note 4**) (cut the tip of the plastic Transfer Pipette according to the size of the embryos).
5. Under the dissecting microscope, using two 1 ml syringes with $29^{1/2}$ G needles; open the embryo along the spinal cord axis, slightly ventral to the spine, by making small tears with one needle while holding the body with the other needle.
6. Locate the primitive lungs in the ventral portion of the embryo, proximal to the incision, below the heart and above the stomach. With the needles, trim away the tissue to free the lungs (*see* **Note 5**). (Embryonic lungs at day 11.5–12.5 in a mouse look like an inverted "Y" with asymmetric "arms" with buds starting to appear on the two bronchi).
7. Under the dissecting microscope, transfer the dissected lungs to a Petri dish containing cold HBSS using a sterile transfer pipette.
8. Prepare 6- or 12-well plates by filling the wells with 1 or 0.5 ml of DMEM/F-12, respectively (*see* **Note 6**) (supplemented with 50 units/ml of penicillin–streptomycin and 0.25 mg/ml Ascorbic Acid).
9. Place each dissected lung from HBSS onto a Nuclepore™ Track-Etched Polycarbonate Membrane Filter (8 μm pore size) using a sterile transfer pipette. Adjust the position of the

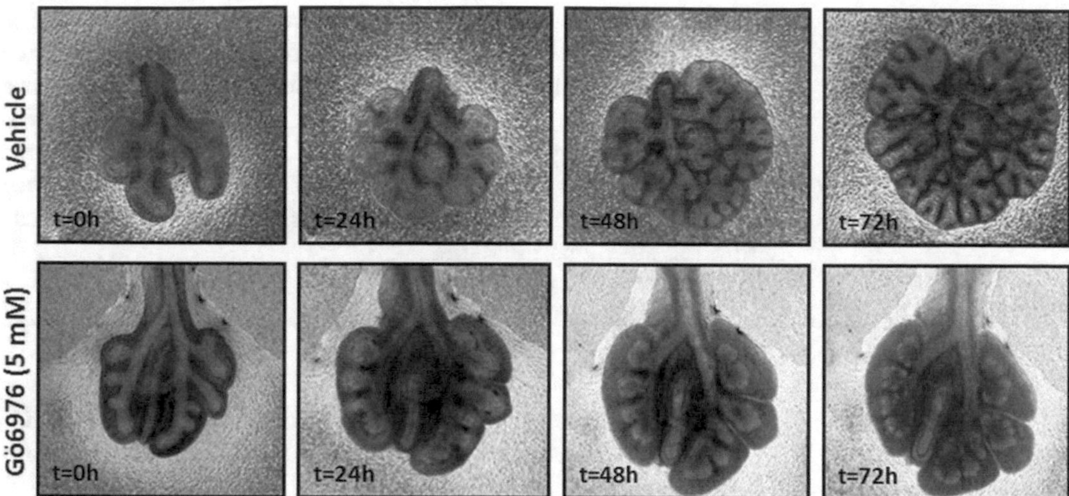

Fig. 1 Effects of Gö6976, a protein kinase C (PKC) inhibitor on in vitro branching morphogenesis of E11.5 lung. Lung explants were treated with Gö6976 (5 μM) or vehicle control for 72 h. Afterwards, the lungs were harvested to examine morphology and number of terminal buds. Examination of the lung explants by light microscopy revealed less branches in the Gö6976-treated rudiments when compared to controls

dissected lungs on the membrane, spreading them evenly on the membrane with their trachea having a flattened and straight position.

10. Gently lay the membranes on the surface of the wells, allowing the membranes to float on top of the media, creating an air–liquid culture setup.
11. Maintain the culture under optimal humidity with an atmosphere of 95% air and 5% CO_2 for the desired culture time.
12. Change the medium every day. Pictures can be taken at different time points. An example of the mouse embryonic lung explants culture and result produced is shown in Fig. 1.

3.2 E17.5 and Adult Mouse Lung Explant Culture

One of the major advantages of early lung embryonic explant culture at an air-liquid interface is that the tissue retains a spherical geometry when the medium is maintained at the correct level. However, due to absence of the entire circulatory system, gaseous diffusion, exchange of nutrients and metabolites are limited to the periphery of the explant. The rate of diffusion limits the size of the explant culture. Thus, at later stages of development, the lung explant culture protocol requires some modifications. E17 or E18 lungs can be chopped into 0.5-mm cubes and cultured submerged in medium [7, 8]. Alternatively, these small pieces of the lungs (0.5-mm cubes) can be cultured on a permeable membrane at an air–liquid interface [9]. Bronchial epithelial cells of explants cultured in a submerged system fail to undergo mucociliary differentiation [10, 11] while small pieces of lungs used under air-liquid interface conditions are not representative of the whole structure of the lungs. To overcome these limitations, air-liquid interface cultures using a lung slice is an another technique that provides a unique approach to investigate in vitro lung physio-pharmacotoxicology [12]. Living lung slices are robust and retain many aspects of the cellular and structural organization of the lung including intrapulmonary airways, arterioles and veins present within the alveoli parenchyma.

4 Methods

4.1 E17.5 and Adult Mouse Lung Explant Culture

1. Euthanize timed-pregnant mice on day E17.5, using a CO_2 chamber, as described previously (*see* **Note 7**). (Clinical death of the animal must be ensured).
2. Spray the abdomen area with 70% ethanol and remove the uterine horns containing embryos from the animal under sterile conditions in a laminar flow hood. Place the embryos in a 50 ml falcon tube containing precooled Hanks Balanced Salt Solution (HBSS) on ice.
3. Place the uterus in a Petri dish containing cold HBSS under the stereoscopic dissecting microscope and release the embryos from the uterus by incising the uterine wall using sterile spring scissors.
4. Transfer the embryos using sterile Dumont forceps into a new Petri dish containing cold HBSS and place it on ice.
5. Remove the head of the embryo and open the chest cavity using Noyes spring scissor. The lung with three lobs can be easily visualized.
6. Under the dissecting microscope, transfer the lungs one by one in to a petri dish containing cold HBSS using sterile Dumont forceps.

4.2 Agarose Embedding (Prepare 30 min before Dissections)

1. Prepare approximately 25 ml of 4% (w/v) low melt agarose in HBSS with glucose for embedding all lobes into small rectangular blocks (*see* **Note 8**). To prepare the agarose: slowly add 1 g of low melting point agarose to 25 ml HBSS, while stirring rapidly with a magnetic bar. Remove the stir bar and microwave until boiling to dissolve the agarose. Add glucose (5 mg/ml) and transfer the mixture to a 50 ml tube on a heat block and maintain the temperature above 40 °C to avoid gelling.
2. Set up the vibratome and the dissection area by sterilizing blades, weighing spatulas, and gather embedding molds, scalpel, fine paintbrush, small scissors, forceps, and a small ice bucket with a sterile petri dish for collecting sections.

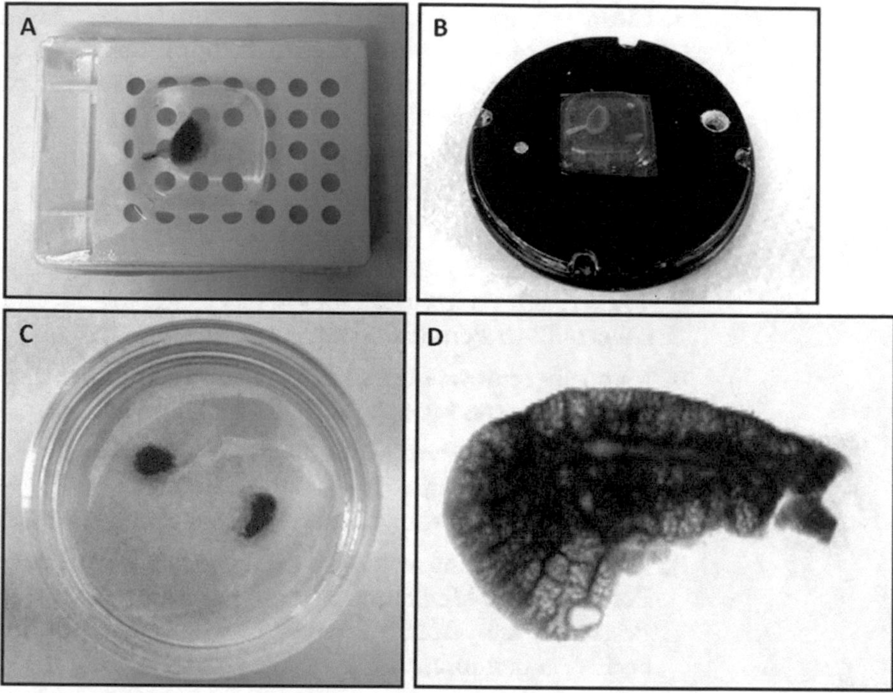

Fig. 2 Agarose Embedding and Lung Slicing. **(a)** A single lobe of fetal lung is embedded in low-melting-point agarose in the mold. **(b)** Agarose block is removed from the mold and fixed onto the specimen mounting disk using adhesive mounting material. **(c)** Two lung slices collected in a Petri dish containing cold HBSS. **(d)** Representative lung slice under stereoscopic dissecting microscope

3. Working in ice-cold HBSS buffer, dissect the lungs at the end of each lobar bronchi to detach each lobe using small scissors.
4. Pat dry each lobe using absorbent sheets to remove excess HBSS. Place each lobe in small metal base molds with at least 3 mm of 4% (w/v) low melting point agarose surrounding the tissue from at the edges. Orient each lobe using forceps by positioning the lobe's largest flat edge at the surface of the metal mold facing the experimenter (Fig. 2a) (*see* **Note 9**). This edge will be fixed to the specimen plate for vibratome sectioning. Place the embedding molds on ice to allow the agarose to solidify.
5. Allow the blocks to harden on ice for at least 5 min prior to sectioning with the vibratome.

4.3 Lung Slicing and Culture

1. Setup the vibratome by filling the sectioning chamber with ice-cold HBSS and maintain a cold temperature throughout sectioning with the surrounding ice bath.
2. Remove the agarose block from the mold and trim down excess agarose surrounding the lobes, while keeping approximately 3 mm from the edge of the tissue.

3. Clean and dry the surface of the specimen mounting disk. Fix the tissue block onto the specimen mounting disk using sufficient adhesive mounting material and submerge the plate into the HBSS-filled sectioning chamber (Fig. 2b).
4. Setup sectioning at 350 μm and slice the agarose block to obtain 350 mm thick slices.
5. Collect sections with spatulas into DMEM/F-12 supplemented with 50 units/ml of Penicillin-Streptomycin in a sterile Petri dish on ice. Fig. 2c shows an example of the lung slices collected in a Petri dish containing cold HBSS.
6. Manually remove the excess agarose from sliced tissue using small scissors and use the slice for air-liquid culture.
7. Prepare 6- or 12-well plates by filling wells with 1 or 0.5 ml of DMEM/F-12, respectively (supplemented with 50 units/ml of penicillin–streptomycin and 0.25 mg/ml Ascorbic Acid).
8. Transfer each lung section onto a Nuclepore™ Track-Etched Polycarbonate Membrane Filter (8 μm pore size) into an empty Petri dish using sterile forceps. Ensure lung slices are spread evenly on membrane.
9. Gently lay the membranes onto the surface of media, allowing the membrane to float on top of media, creating an air-liquid culture. Slices must be in the incubator within 2 h following initial dissection.
10. Maintain the culture under optimal humidity with an atmosphere of 95% air and 5% CO_2 for the desired culture time.
11. Change the medium every day. Slices can be maintained in vitro for 72 h, depending on treatment. Pictures can be taken at different time points, or lung slices can be harvested and examined via histology. An example of the lung slice (E17.5) culture and result produced is shown in Fig. 3.

5 Notes

1. Place the animal in a chamber and expose the animal to 100% CO_2 at a fill rate of 10–30% of chamber volume per minute. Following unconsciousness, observe the animal for faded eye color and lack of respiration. When both are observed, maintain CO_2 filling for 1–2 min and then remove animal from chamber.
2. The abdomen area of the animal could be disinfected with 70% ethanol.
3. Use a sterile spring scissor to incise the uterine wall and release the embryos from the uterus. A timed-pregnant female yields on average 8–10 embryos.

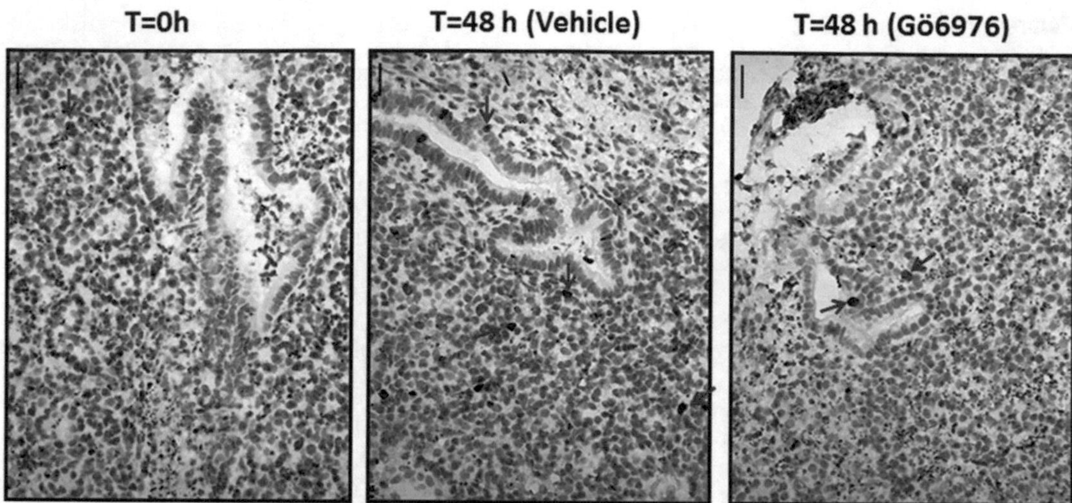

Fig. 3 Effects of Gö6976, a protein kinase C (PKC) inhibitor on mouse fetal lung proliferation. Fetal lungs (E17.5) were sliced and cultured in air-liquid interface. Explants were treated with bromodeoxyuridine (BrdU) reagent, an indicator of cell proliferation, at 10 μM for 6 h, followed by treatment with Gö6976 (5 μM) or vehicle control for 48 h. Afterward, the lung slices were harvested, fixed in 4% paraformaldehyde and sectioned. Immunohistochemistry was carried out on sections with a monoclonal antibody specific for BrdU. Immunohistochemistry of the slices for BrdU, revealed reduced numbers of BrdU positive cells (Red arrows, dark brown nuclear staining) in the Gö6976-treated lung slices when compared to control

4. Cut the tip of a disposable plastic Transfer Pipette according to the size of the embryos. Transfer one embryo at a time by gently squeezing the plastic Transfer Pipette with your fingers.
5. Embryonic lungs at day 11.5–12.5 in a mouse look like an inverted "Y" with asymmetric "arms" with buds starting to appear on the two bronchi.
6. Prior to use, supplement the DMEM/F-12 medium with 50 units/ml Penicillin-Streptomycin and 0.25 mg/ml Ascorbic Acid.
7. Death of the animal must be ensured by an appropriate method, such as ascertaining cardiac and respiratory arrest or noting fixed and dilated pupils.
8. To prepare the agarose: slowly add 1 g of low melt agarose to 25 ml HBSS, while stirring rapidly with a magnetic bar. Remove the stir bar and microwave until boiling to dissolve the agarose. Add glucose (5 mg/ml) and transfer the mixture to a 50 ml tube on a heat block and maintain the temperature above 40 °C to avoid gelling.
9. This edge will be fixed to the specimen plate for vibratome sectioning.

References

1. Randall KJ, Turton J, Foster JR (2011) Explant culture of gastrointestinal tissue: a review of methods and applications. Cell Biol Toxicol 27(4):267–284
2. Narhi K, Thesleff I (2010) Explant culture of embryonic craniofacial tissues: analyzing effects of signaling molecules on gene expression. Methods Mol Biol 666:253–267
3. Narhi K (2017) Embryonic explant culture: studying effects of regulatory molecules on gene expression in craniofacial tissues. Methods Mol Biol 1537:367–380
4. Trowell OA (1959) The culture of mature organs in a synthetic medium. Exp Cell Res 16(1):118–147
5. Nicholas B, Staples KJ, Moese S, Meldrum E, Ward J, Dennison P, Havelock T, Hinks TS, Amer K, Woo E, Chamberlain M, Singh N, North M, Pink S, Wilkinson TM, Djukanovic R (2015) A novel lung explant model for the ex vivo study of efficacy and mechanisms of anti-influenza drugs. J Immunol 194(12): 6144–6154
6. Del Moral PM, Warburton D (2010) Explant culture of mouse embryonic whole lung, isolated epithelium, or mesenchyme under chemically defined conditions as a system to evaluate the molecular mechanism of branching morphogenesis and cellular differentiation. Methods Mol Biol 633:71–79
7. Shan L, Emanuel RL, Dewald D, Torday JS, Asokanathan N, Wada K, Wada E, Sunday ME (2004) Bombesin-like peptide receptor gene expression, regulation, and function in fetal murine lung. Am J Physiol Lung Cell Mol Physiol 286(1):L165–L173
8. Sunday ME, Hua J, Dai HB, Nusrat A, Torday JS (1990) Bombesin increases fetal lung growth and maturation in utero and in organ culture. Am J Respir Cell Mol Biol 3(3):199–205
9. Prince LS, Okoh VO, Moninger TO, Matalon S (2004) Lipopolysaccharide increases alveolar type II cell number in fetal mouse lungs through Toll-like receptor 4 and NF-kappaB. Am J Physiol Lung Cell Mol Physiol 287(5): L999–1006
10. Prytherch Z, Job C, Marshall H, Oreffo V, Foster M, BeruBe K (2011) Tissue-Specific stem cell differentiation in an in vitro airway model. Macromol Biosci 11(11): 1467–1477
11. Shojaie S, Ermini L, Ackerley C, Wang J, Chin S, Yeganeh B, Bilodeau M, Sambi M, Rogers I, Rossant J, Bear CE, Post M (2015) Acellular lung scaffolds direct differentiation of endoderm to functional airway epithelial cells: requirement of matrix-bound HS proteoglycans. Stem Cell Reports 4(3):419–430
12. Morin JP, Baste JM, Gay A, Crochemore C, Corbiere C, Monteil C (2013) Precision cut lung slices as an efficient tool for in vitro lung physio-pharmacotoxicology studies. Xenobiotica 43(1):63–72

Check for updates

Chapter 9

Isolating Embryonic Cardiac Progenitors and Cardiac Myocytes by Fluorescence-Activated Cell Sorting

Abdalla Ahmed and Paul Delgado-Olguin

Abstract

Isolation of highly purified populations of embryonic cardiomyocytes enables the study of congenital cardiac phenotypes at the cellular level. Fluorescent-activated cell sorting (FACS) is normally used to isolate fluorescently tagged cells. Here we describe the isolation of differentiating mouse embryonic cardiac progenitors and cardiomyocytes at embryonic day (E) 9.5 and E13.5, respectively by FACS. Over 50,000 differentiating cardiac progenitors and 200,000 cardiomyocytes can be obtained in a single prep using the methods described.

Key words Fluorescent-activated cell sorting, Cardiac progenitor isolation, Cardiomyocyte isolation, Mouse embryo

1 Introduction

Fluorescence activated cell sorting, involving the use of fluorescent molecules to tag cells expressing specific markers [1–3], has become a monumental driving force in isolation of distinct cell populations from heterogeneous samples or tissues. The technique takes advantage of the fact that different cell types express unique markers that distinguish them from the remainder of the cell population. Cells expressing the unique marker, or a combination of markers, can then be tagged through a variety of methods most commonly using (a) fluorescently tagged antibodies recognizing cell surface makers (used for isolation of live cells) [4, 5], (b) mice/cell lines carrying transgenes driving expression of fluorescent proteins (e.g., GFP) in specific tissues (also used for isolation of live cells) [6, 7], and c) antibodies against nuclear proteins (used to isolate nuclei from lysed cells) [8, 9].

Isolation of cardiomyocytes from the embryonic heart is of particular importance as recent studies have shown that changes in embryonic cardiomyocytes can predispose to adult disease [10, 11]. Moreover, advances in the field have elucidated how different cardiac

Paul Delgado-Olguin (ed.), *Mouse Embryogenesis: Methods and Protocols*, Methods in Molecular Biology, vol. 1752, https://doi.org/10.1007/978-1-4939-7714-7_9, © Springer Science+Business Media, LLC, part of Springer Nature 2018

diseases may arise from defects in specific cell types in the heart, highlighting the importance of studying the various cell populations of the heart [12–14]. Isolation of pure cell populations from the heart is challenging owing to their heterogeneity. Cardiac cell types include fibroblasts, cardiomyocytes, endothelial cells, smooth muscle, and epicardial cells. In addition, the various cell markers express at different levels and change their specificity throughout development. Hence, different cell types express different markers at different stages of embryogenesis. Nk2 homeobox 5, or Nkx2–5, is a transcription factor and a marker of cardiac progenitors. *Nkx2–5* expression is restricted to the developing heart between E7.5 and E9.5 and labels a mixed population of differentiating and differentiated cardiomyocytes by E9.5 [15]. Using transgenic mice expressing eGFP driven by the *Nkx2–5* promoter, live differentiating cardiac progenitor populations can be isolated from whole embryos between E7.5 and E9.5 [16]. In addition, using a fluorescently tagged VCAM1 antibody, live cardiomyocyte-enriched populations can be isolated from whole embryonic hearts [4]. VCAM1 is a cell surface protein and a marker of differentiated embryonic cardiomyocytes [4]. It is specifically expressed in cardiomyocytes between E11.5 and E13.5 [4]. Here we describe a method for efficient isolation of differentiating cardiac progenitors from whole embryos at E9.5 using transgenic mice, and cardiomyocytes from whole hearts at E11.5 and E13.5 by FACS using an anti VCAM1 antibody.

2 Materials

Prepare all solutions in autoclaved double distilled water. Consult appropriate Material Safety Data Sheets for proper handling and disposal of materials.

2.1 Isolating Cardiac Progenitor Cells from E7.5–9.5 Mouse Embryos

2.1.1 Reagents

Autoclaved double distilled water.

1 × PBS (137 mM NaCl, 2.7 mM KCl, 8 mM Na_2HPO_4, and 2 mM KH_2PO_4 pH 7.4).

Red Blood Cell (RBC) lysis buffer (10×) (Miltenyi Biotec).

Fetal bovine serum (FBS) (*see* **Note 1**).

DMEM containing high glucose, sodium pyruvate, sodium bicarbonate.

1000× propidium iodide (1 mg/ml).

1× TrypLE™ Express.

Cell Culture Medium: 10% FBS in DMEM supplement for a final concertation of 1× nonessential amino acids (Wisent Inc) and 1% streptomycin–penicillin.

Cell Sorting Medium: 1% FBS in DMEM 1 mM EDTA.

Digestion Inhibitor Media: 10% FBS in DMEM.

2.1.2 Equipment

Dissection tweezers.

6 cm petri dishes.

Dissecting microscope.

Florescence microscope capable of detecting GFP.

5 ml Round Bottom Polystyrene Test Tube, with 35 μm Cell Strainer Snap Cap.

5 ml round-bottom polypropylene tubes.

Eppendorf™ Snap-Cap 2 ml round bottom Microcentrifuge Safe-Lock Tubes, sterile.

Disposable plastic pipette. Cut the tip off to allow suction of individual whole embryos.

Water bath set at 37 °C.

Centrifuge.

Refrigerated centrifuge set to 4C.

Flow Cytometer capable of detecting eGFP and propidium iodide.

2.1.3 Mouse Line

B6;129P2-Tg(NKX2–5-EmGFP)2Conk/Mmnc. Other lines expressing GFP in cardiac progenitor cells or cardiomyocytes can be used.

2.2 Isolating Cardiomyocytes from E11.5–13.5 Mouse Hearts

2.2.1 Reagents

Autoclaved double distilled water.

1 × PBS (137 mM NaCl, 2.7 mM KCl, 8 mM Na_2HPO_4, and 2 mM KH_2PO_4 pH 7.4).

Red Blood Cell (RBC) lysis buffer (10×).

Fetal Bovine Serum (FBS) (*see* **Note 1**).

DMEM containing high glucose, sodium pyruvate, sodium bicarbonate.

1000× propidium iodide (1 mg/ml).

Neonatal heart dissociation kit (*see* **Note 2**).

Cell Culture Medium: 10% FBS in DMEM supplement for a final concertation of 1× nonessential amino acids and 1% streptomycin–penicillin.

Cell Sorting Medium: 1% FBS in DMEM 1 mM EDTA.

VCAM1 antibody-conjugated with Alexa 647, anti-mouse.

2.2.2 Equipment

Dissection Tweezers.

6 cm petri dishes.

Dissecting microscope.

5 ml Round Bottom Polystyrene Test Tube, with 35 μm Cell Strainer Snap Cap.

5 ml round-bottom polypropylene tubes.

Eppendorf™ Snap-Cap 2 ml round bottom Microcentrifuge Safe-Lock™ Tubes, sterile.

Water bath set at 37 °C.

Centrifuge.

Refrigerated centrifuge set to 4 °C.

Flow Cytometer capable of detecting Alexa 647 and propidium iodide.

2.2.3 Mouse Line

C57BL/6 J (*see* **Note 3**).

3 Methods

3.1 Isolating Cardiac Progenitor Cells from E7.5–9.5 Mouse Embryos

Prior to dissection keep PBS on ice for at least 30 min to ensure that it is cold. Set the water bath to 37 °C.

1. Animal procedures must follow Institutional guidelines and must be approved by the corresponding Animal Care Committee. Terminate the pregnant dam through cervical dislocation, immediately dissect the uterus and place it in a 6 cm petri dish containing cold 1× PBS.
2. Dissect whole embryos from the uterus under a dissection microscope and remove all extra-embryonic tissues ensuring complete removal of both yolk sac and amniotic sac.
3. Use a disposable plastic pipette to transfer embryos to a new petri dish containing cold PBS. This improves visualization and facilitates counting of somites.
4. Count somites to ensure embryos are properly staged and discard any embryos that are not at the appropriate stage (E9.5 = 21–25 somites) (*see* **Note 4**).
5. Check the embryos under a fluorescence microscope to identify GFP positive and GFP negative embryos. Ensure you keep one GFP negative embryo to serve as a negative control for setting up the cell sorting gate.
6. Transfer the embryos to 2 ml round bottom tubes using a plastic pipette. Gently remove all PBS carried over using a P200 pipette.
7. If digesting individual embryos (final cell yield approx. 5000 cardiac progenitors/embryo):
   - Add 300 μl TrypLE™ Express to each tube containing an individual embryo.
   - Incubate at 37 °C for 10 min, pipetting up and down every 3 min using P200 tips to ensure complete dissociation of the embryo to a single cell suspension.

If digesting an entire litter/10 embryos (final cell yield approx. 50,000 cardiac progenitors per litter).

- Add 700 μl TrypLE™ Express to the tube containing all the embryos (up to 15 embryos).
- Incubate at 37 °C for 15 min and pipette up and down every 5 min to using P200 tips to ensure complete dissociation of the embryos to single cell suspensions.

8. For single embryo dissociation: Add 30 μl FBS and 300 μl 10% FBS DMEM.

   For whole litter dissociation: Add 70 μl FBS and 700 μl 10% FBS DMEM.

   Mix gently by inversion.

9. Centrifuge at 270 RCF for 5 min at room temperature to pellet the cells.
10. Meanwhile, prepare the 1× RBC lysis buffer by diluting the 10× RBC stock buffer in ddH_20. Keep the lysis buffer at room temperature.
11. Remove the supernatant from the cell pellet.
12. Resuspend the cell pellet in 2 ml of 1× RBC lysis buffer.
13. Incubate at room temperature for 5 min with gentle mixing by inversion every 30 s.
14. Centrifuge at 270 RCF for 3 min.
    - The cell pellet should be a white color. If the pellet looks red, resuspend the pellet in the same volume of 1× RBC lysis solution and incubate at room temperature for another 3 min.
15. Remove the supernatant from the cell pellet.
16. Resuspend in 300 μl cell sorting media (1% FBS DMEM 1 mM EDTA) (*see* **Note 5**).
17. Add propidium iodide to each tube to a final concentration of 1 μg/ml (i.e., 0.3 μl of stock solution in 300 μl). Mix by gentle pipetting.
18. Filter Cells by pipetting them through the 35 μm cell strainer into a 5 ml polypropylene test tube.
19. Place the resuspended cells on ice and proceed to cell sorting (*see* **Note 6**). If conducted properly, the expected cell viability is approximately 85%. GFP positive cells account for approx. 2–5% of viable cells (Fig. 1).
20. Collect cells in 300 μl cell culture media and proceed with desired downstream analysis.

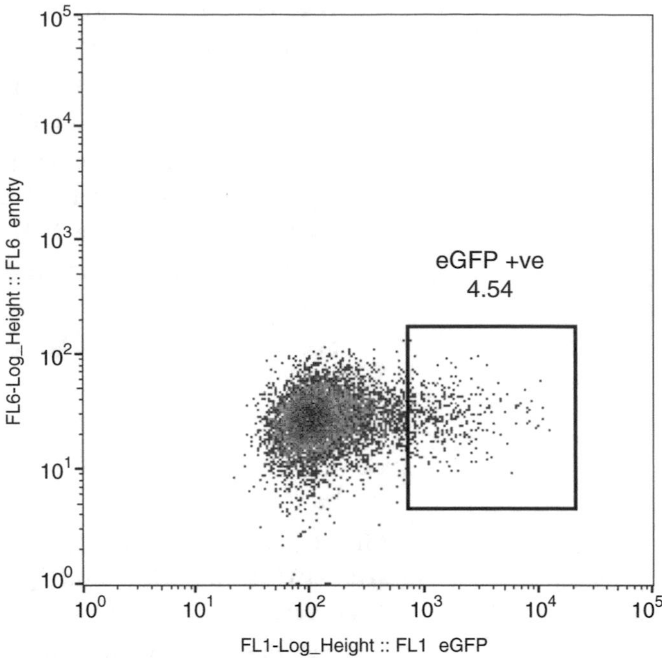

Fig. 1 Sorting gate for isolation of E9.5 cardiac progenitors. Cardiac progenitors are represented in the GFP positive population and represent <5% of the total cells in the embryo

3.2 Isolating Cardiomyocytes from E11.5–13.5 Mouse Hearts

Prior to dissection keep PBS on ice for at least 30 min to ensure it is cold. Set the water bath to 37 °C.

1. Animal procedures must follow Institutional guidelines and must be approved by the corresponding Animal Care Committee. Terminate the pregnant dam through cervical dislocation and immediately extract the uterus and place it in a 6 cm petri dish containing cold 1× PBS.
2. Dissect whole embryos from the uterus under a dissection microscope.
3. Extract the embryonic hearts (E11.5–13.5) and place them in ice cold 1×PBS (*see* **Note 7**).
4. Break apart the ventricles and rinse in 1×PBS to remove some of the blood from the ventricles.
5. Dissociate the ventricles using the neonatal heart dissociation kit:
   (a) Place the hearts in a 2 ml round bottom tube.
   (b) Remove PBS completely using P200 pipette.
   (c) Add 460 μl Solution 2 + 12.5 μl Solution 1 + 5 μl Solution 3 + 2.5 μl Solution 4 + 20 μl Solution 5.
   (d) Incubate in water bath at 37 °C pipetting up and down vigorously using a P200 pipette every 4–6 min. Digest for

a total time of 15 min (or until hearts are completely dissociated).

6. Centrifuge at 270 RCF for 5 min at room temperature to pellet the cells.
7. Meanwhile, prepare the 1 × RBC lysis buffer by diluting the 10× stock buffer in water. Keep at room temperature.
8. Remove the supernatant from the cell pellet.
9. Resuspend the cell pellet in 2 ml of 1× RBC lysis buffer.
10. Incubate at room temperature for 5 min and with gentle mixing every 30 s.
11. Remove 30–50 μl of the solution and set them in a separate 2 ml round bottom tube. This will serve as the negative control for setting up the sorting gate.
12. Centrifuge both tubes at 270RCF for 3 min.
    - The cell pellet should be of white color. If the color is red, resuspend the pellet in the same volume of 1 × RBC lysis solution and incubate at room temperature for another 3 min).
13. Meanwhile prepare the antibody solution by adding 50 μl FBS to 945 μl 1× PBS and 5 μl VCAM1 antibody (0.2 mg/ml). Mix on vortex and place on ice. Prepare a similar solution without the antibody for the negative control.
14. Remove the supernatant from the cell pellets.
15. Resuspend the cell pellet in 1 ml 5% FBS 1× PBS with 1 μg/ml VCAM1 antibody (prepared in **step 14**). Resuspend the negative control in the solution without the antibody (prepared in **step 14**).
16. Place both tubes containing cells under ice for 15 min and cover the ice bucket from light to prevent photobleaching.
17. Centrifuge at 270RCF for 5 min at 4 °C to pellet the cells. Remove the supernatant from the cell pellet.
18. Resuspend the pellet in 400 μl cell sorting medium (1%FBS DMEM 1mMEDTA).
19. Add propidium iodide to each tube to a final concentration of 1 μg/ml (i.e., 0.4 μl of stock solution per 400 μl). Mix by gently pipetting.
20. Filter Cells by pipetting them through the 35 μm cell strainer into a 5 ml polypropylene test tube.
21. Place cells on ice and proceed to cell sorting (*see* **Note 6**). If conducted properly, the expected cell viability should be >95%. VCAM1 positive cells account for approximately 40% of viable cells (Fig. 2) and include >85% cardiomyocytes (Fig. 3).
22. Collect cells in 300 μl cell culture media and proceed with desired downstream analysis.

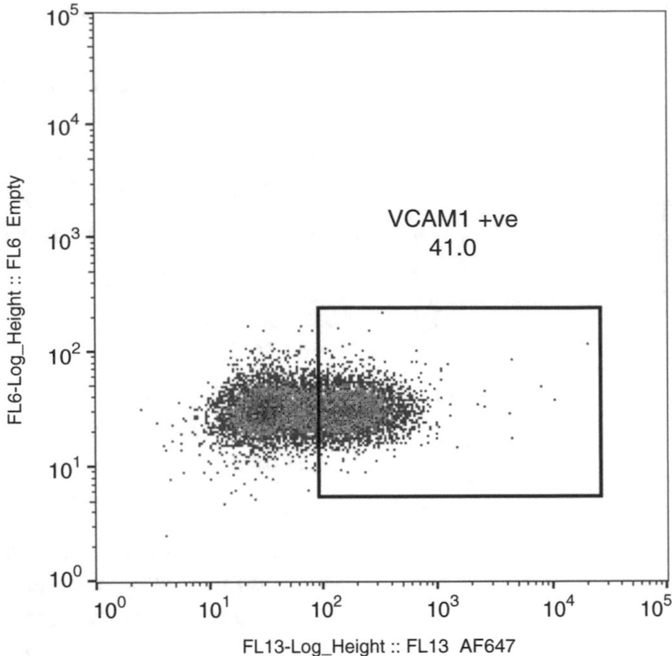

Fig. 2 Sorting gate for isolation of E13.5 cardiomyocytes. Cardiomyocytes are represented in the Alexa 647/VCAM1 positive population and represent ~40% of the total cells in the heart

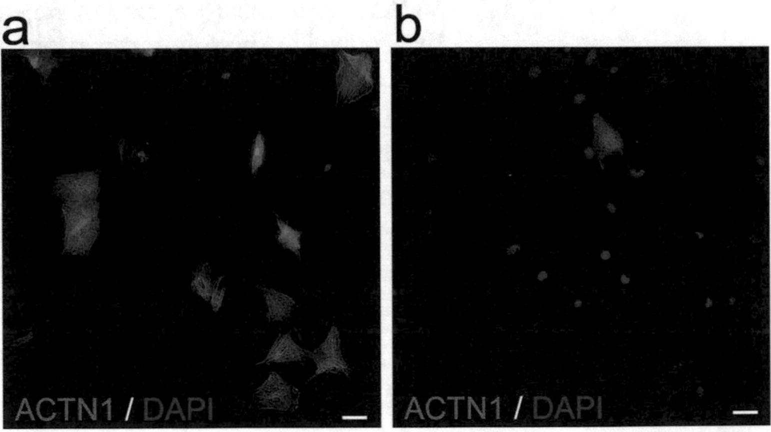

Fig. 3 Immunofluorescence images against sarcomeric marker a-actinin (ACTN1). (**a**) VCAM1 positive cells are enriched for cardiomyocytes. (**b**) VCAM1 negative cells are depleted from cardiomyocytes. Nuclei are stained with DAPI (blue)

4 Notes

1. Ensure that you use FBS from the same origin for all experiments as serum from different animals may have different downstream effects on the cells.
2. We follow the digestion protocol provided in the Neonatal heart dissociation kit, however using one tenth of the recommended volumes for digestion is sufficient for digestion of embryonic hearts.
3. This protocol was optimized using C57BL/6 J genetic background. Other genetic backgrounds (such as CD-1) which have bigger hearts may require longer digestion and/or RBC lysis times.
4. It is common to see variation in somite numbers within the same litter at these early stages of development hence it is critical to count somites and collect cells from developmentally matched embryos to reduce variability in final results of the experiments.
5. Different facilities and sorting machines may require different cell dilution, consult with your flow cytometry facility or machine manual for optimal cell dilutions.
6. When setting up the sorting gates, ensure to include cells with high forward and side scatter heights as cardiomyocytes tend to be more complex. Ensure sorting is extremely slow (<700 event per sec) with a very low differential pressure (<0.5) as embryonic cardiac progenitors and cardiomyocytes are extremely sensitive to high pressures.
7. We recommend separating the atria from the ventricles at this point as they contain functionally distinct cardiomyocyte cell populations.

Acknowledgment

We thank The SickKids-UHN Flow Cytometry Facility for help with FACS, and The Centre for Phenogenomics (TCP) for mouse husbandry. This work was supported by the Heart and Stroke Foundation of Canada (G-17-0018613), the Natural Sciences and Engineering Research Council of Canada (NSERC) (500865), the Canadian Institutes of Health Research (CIHR) (PJT-149046), and Operational Funds from the Hospital for Sick Children to P.D.-O.

References

1. Loken MR, Herzenber LA (1975) Analysis of cell populations with a fluorescence-activated cell sorter. Ann N Y Acad Sci 254:163–171
2. Han Y, Gu Y, Zhang AC, Lo YH (2016) Review: imaging technologies for flow cytometry. Lab Chip 16(24):4639–4647
3. Herzenberg LA, Parks D, Sahaf B, Perez O, Roederer M, Herzenberg LA (2002) The history and future of the fluorescence activated cell sorter and flow cytometry: a view from Stanford. Clin Chem 48(10):1819–1827
4. Pontén A, Walsh S, Malan D, Xian X, Schéele S, Tarnawski L, Fleischmann BK, Jovinge S (2013) FACS-based isolation, propagation and characterization of mouse embryonic cardiomyocytes based on VCAM-1 surface marker expression. PLoS One 8(12):e82403
5. Dubois NC, Craft AM, Sharma P, Elliott DA, Stanley EG, Elefanty AG, Gramolini A, Keller G (2011) SIRPA is a specific cell-surface marker for isolating cardiomyocytes derived from human pluripotent stem cells. Nat Biotechnol 29(11):1011–1018
6. Gantz JA, Palpant NJ, Welikson RE, Hauschka SD, Murry CE, Laflamme MA (2012) Targeted genomic integration of a selectable floxed dual fluorescence reporter in human embryonic stem cells. PLoS One 7(10):e46971
7. Lee MY, Sun B, Schliffke S, Yue Z, Ye M, Paavola J, Bozkulak EC, Amos PJ, Ren Y, Ju R, Jung YW, Ge X, Yue L, Ehrlich BE, Qyang Y (2012) Derivation of functional ventricular cardiomyocytes using endogenous promoter sequence from murine embryonic stem cells. Stem Cell Res 1:49–57
8. Bergmann O, Zdunek S, Alkass K, Druid H, Bernard S, Frisén J (2011) Identification of cardiomyocyte nuclei and assessment of ploidy for the analysis of cell turnover. Exp Cell Res 317(2):188–194
9. Gilsbach R, Preissl S, Grüning BA, Schnick T, Burger L, Benes V, Würch A, Bönisch U, Günther S, Backofen R, Fleischmann BK, Schübeler D, Hein L (2014) Dynamic DNA methylation orchestrates cardiomyocyte development, maturation and disease. Nat Commun 5:5288
10. Delgado-Olguín P, Huang Y, Li X, Christodoulou D, Seidman CE, Seidman JG, Tarakhovsky A, Bruneau BG (2012) Epigenetic repression of cardiac progenitor gene expression by Ezh2 is required for postnatal cardiac homeostasis. Nat Genet 44(3):343–347
11. Sah R, Mesirca P, Mason X, Gibson W, Bates-Withers C, Van den Boogert M, Chaudhuri D, Pu WT, Mangoni ME, Clapham DE (2013) Timing of myocardial trpm7 deletion during cardiogenesis variably disrupts adult ventricular function, conduction, and repolarization. Circulation 128(2):101–114
12. Hamdani N, Kooij V, van Dijk S, Merkus D, Paulus WJ, Remedios CD, Duncker DJ, Stienen GJ, van der Velden J (2008) Sarcomeric dysfunction in heart failure. Cardiovasc Res 77(4):649–658
13. Song L, Zhao M, Wu B, Zhou B, Wang Q, Jiao K (2011) Cell autonomous requirement of endocardial Smad4 during atrioventricular cushion development in mouse embryos. Dev Dyn 240(1):211–220
14. Fan D, Takawale A, Lee J, Kassiri Z (2012) Cardiac fibroblasts, fibrosis and extracellular matrix remodeling in heart disease. Fibrogenesis Tissue Repair 5(1):15
15. Lints TJ, Parsons LM, Hartley L, Lyons I, Harvey RP (1993) Nkx-2.5: a novel murine homeobox gene expressed in early heart progenitor cells and their myogenic descendants. Development 119(2):419–431
16. Hsiao EC, Yoshinaga Y, Nguyen TD, Musone SL, Kim JE, Swinton P, Espineda I, Manalac C, deJong PJ, Conklin BR (2008) Marking embryonic stem cells with a 2A self-cleaving peptide: a NKX2-5 emerald GFP BAC reporter. PLoS One 3(7):e2532

Check for updates

Chapter 10

Isolation and Culture of Mouse Placental Endothelial Cells

Lijun Chi and Paul Delgado-Olguin

Abstract

Isolation and culture of endothelial cells (ECs) is a useful tool to study the cellular processes involved in vascular development and vascular maturation. In this chapter, we describe a method to isolate and culture endothelial cells from placentae. This method takes advantage of two transgenes: $ROSA26^{mT/mG}$, which drives the expression of *GFP* upon Cre-mediated recombination, and *Tie2-Cre*, which expresses Cre driven by the *Tie2* promoter in endothelial progenitors and their descendants. GFP-expressing endothelial cells are isolated through fluorescence-activated cell sorting (FACS). The sorted cells express the endothelial marker CD31. This method can be used to study the morphological and physiological properties of placental endothelial cells in mice carrying mutations affecting vascular development.

Key words Mouse placenta, Endothelial cells, Fluorescence-activated cell sorting, Primary cell culture

1 Introduction

Endothelial cells (ECs) drive the establishment and organization of vascular networks [1, 2] and form a single-cell layer outlining the blood vessel lumen. ECs participate in numerous processes required for embryogenesis and organ homeostasis including establishment of a permeable barrier and vascular structure maintenance, and mediate the response of shear stress caused by blood flow [3]. Significant efforts have focused at uncovering the functions of endothelial cells in vascular development in the embryo proper, and on harnessing their potential in experimental tissue regeneration strategies. In contrast, the processes controlling endothelial cell development and formation of the placental vascular network of embryonic origin, known as the labyrinth in mice, are poorly understood. Based on published protocols used for isolation and culture of mouse vascular, lung, heart, and human umbilical ECs (HUVECs) [4–8], we have developed a method for isolation and culture of labyrinth endothelial cells from placenta at embryonic day (E) 12.5.

Paul Delgado-Olguin (ed.), *Mouse Embryogenesis: Methods and Protocols*, Methods in Molecular Biology, vol. 1752, https://doi.org/10.1007/978-1-4939-7714-7_10, © Springer Science+Business Media, LLC, part of Springer Nature 2018

This method takes advantage of two transgenes: *ROSA26*$^{mT/mG}$, and *Tie2-Cre*. Mice carrying the *ROSA26*$^{mT/mG}$ transgene constitutively express membrane tomato GFP, and conditionally express membrane GFP upon Cre-mediated recombination. In combination with tissue-specific expression of the Cre recombinase, this transgene can be used to fluorescently label specific cell populations [9]. Thus, mice carrying *ROSA26*$^{mT/mG}$, and *Tie2-Cre* [10] constitutively express membrane tomato GFP, and membrane GFP exclusively in endothelial progenitors and their descendants. These cells, which are positive for the endothelial cell marker CD31 can be then isolated by FACS and cultured for downstream analysis.

2 Materials

Mouse lines: *ROSA26*$^{mT/mG}$ [9] and *Tie2-Cre* [10]. All reagents and solutions are prepared in autoclaved *Milli-Q* water (filtered through a 0.22 μm membrane) at room temperature. Sterilize dissection tools in autoclave.

2.1 Equipment

1. Dissection microscope (e.g., Leica MZ10F).
2. 37 °C water bath.
3. Tissue culture incubator at 37 °C supplied with 5% CO_2.
4. Biosafety cabinet (BSC).
5. Fluorescence-activated cell sorter-MoFlo Astrios BRVY.
6. Hemocytometer.
7. Inverted or upright microscope.
8. Phase contract microscope.
9. Confocal microscope with proper lasers to stimulate GFP and DAPI fluorescence.

2.2 Dissection of Placenta

1. Pregnant mouse (E12.5) (*see* **Note 1**).
2. Straight dissecting forceps (110 mm).
3. Curved forceps (120 mm).
4. Straight scissors (140 mm).
5. Tissue culture dish (60 mm).

2.3 Tubes and Cell Culture Plates

1. 12-well cell culture plates.
2. 96-well cell culture plates.
3. 50 ml conical tubes.
4. 15 ml conical tubes.
5. 5 ml polypropylene round-bottom tubes.
6. 1.5 ml microcentrifuge tubes.
7. 70 μm cell strainer.

8. Chamber tissue culture glass slide.
9. Cover glass.

2.4 Reagents, Solutions, and Antibodies

1. 70% ethanol.
2. Phosphate-Buffered Saline (10 × PBS, pH 7.4) 1 l. To 700 ml of water add: 80 g NaCl, 2 g KCl, 14.4 g Na_2HPO_4, and 2.4 g of KH_2PO_4. Bring final volume to 1 l and adjust pH to 7.4 with HCl. Sterilize in autoclave.
3. TrypLE™ Express.
4. Red blood cell lysis solution (10×) (Miltenyi Biotec) (*see* **Note 2**).
5. Propidium iodide (PI) at 2 μg/ml. Prepare stock at 1 mg/ml (*see* **Note 3**)
6. 0.1% gelatin in 1× PBS pH 7.4. Store at 4 °C (*see* **Note 4**).
7. High glucose Dulbecco's modified Eagle's medium (DMEM). Store at 4 °C.
8. Fetal bovine serum (FBS). Inactivate serum by incubating in a water bath at 55 °C for 60 min. Aliquot in 50 ml conical tubes and store at −20 °C.
9. 0.5 M EDTA, pH 8.0, 500 ml. Add 93.05 g of ethylenediaminetetraacetic acid to 400 ml of dH_2O. Adjust the pH to 8.0 with NaOH (~25 ml of 10 N NaOH). Bring volume to 500 ml with dH_2O. Sterilize in autoclave.
10. Sorting Medium 50 ml: Add 500 μl FBS (Final concentration 1%) and 100 μl of 0.5 M EDTA to 49.4 ml of Dulbecco's modified Eagle's medium (DMEM).
11. Cell collection medium 50 ml: Add 5 ml of FBS to 45 ml Dulbecco's modified Eagle's medium (DMEM).
12. Endothelial culture medium (ECM) (Commercially available).
13. 10× endothelial cell growth supplement (ECGS) (Commercially available).
14. Antibiotic cocktail. Penicillin 10,000 U/ml + streptomycin 10,000 U/ml.
15. Recombinant human vascular endothelial growth factor (hVEGF) (*see* **Note 5**).
16. Matrigel basement membrane matrix (*see* **Note 6**).
17. Endothelial cell culture medium work solution 5 ml (*see* **Note 7**):

Mix 500 μl of 10× endothelial cell growth supplement (ECMS), 50 μl of penicillin 10,000 U/ml) + streptomycin 10,000 U/ml, 10 μl of hVEGF (50 μg/ml), and 750 μl FBS, then bring volume to 5 ml with endothelial culture medium (ECM).

18. 4% Paraformaldehyde (PFA). Dissolve 20 g of PFA in 500 ml of 1× PBS. Incubate in a water bath at 65 °C in a fume hood, add 3 ml of 10 N NaOH and gently mix until the solution

clears. Aliquot in 10 ml tubes in a fume hood and store for long term at −20 °C. Store aliquot of working solution at 4 °C.

19. Blocking buffer 5 ml: Mix 500 μl of 30% bovine serum albumin (BSA), 250 μl goat serum, and 5 μl Tween 20, and bring the volume to 5 ml with 1× PBS, pH 7.4.
20. Incubation buffer 5 ml: Mix 250 μl of 30% BSA, 125 μl goat serum, and 5 μl Tween 20. Bring the volume to 5 ml with 1× PBS, pH 7.4.
21. Washing buffer 5 ml: Mix 25 μl of 30% BSA, 12.5 μl goat serum, and 5 μl Tween 20. Bring volume to 5 ml with 1× PBS, pH 7.4.
22. Rat Anti-mouse CD31 antibody.
23. GFP antibody (Chicken).
24. Alexa Fluor® 488 goat anti-chicken IgG.
25. Alexa Fluor® 546 goat anti-mouse IgG.
26. DAPI (4′,6-diamidino-2-phenyindole). Dissolve 5 mg in distilled water to 10 mg/ml. The solution will become yellow clear. Protect from light. For long-term storage, aliquot and freeze at −20 °C (*see* **Note 8**).
27. Vectashield HardSet mounting medium.

3 Methods

To prevent contamination of the cell cultures, all solutions, tools, and dishes must be sterile. Animal procedures must follow Institutional guidelines and be approved by the corresponding Animal Care Committee. Euthanize pregnant mice at embryonic days (E) E11.5 to E13.5.

3.1 Labyrinth Tissue Dissection

1. Dissect the placenta form *Rosa26*$^{mT/mG}$;*Tie2-Cre* embryos in precooled 1× PBS in a 60 mm tissue culture dish.
2. Dissect the labyrinth (Dotted line in Fig. 1a) from the placenta.
3. Using the Straight dissecting forceps separate the labyrinth tissue into small pieces, and gently rinse them in 1× PBS three times to remove blood (*see* **Note 9**).

3.2 Dissociation of Labyrinth and Endothelial Cell for FACS

1. Transfer the labyrinth pieces into a 12-well plate, add the appropriate volume of TrypLE™ Express solution (*see* **Note 10**) depending on the amount of tissue to be disassociated.
2. Incubate at 37 °C in a cell culture incubator for 30 min. To release more cells, gently agitate the sample by pipetting up and down every 10 min using 1000 μl tips.

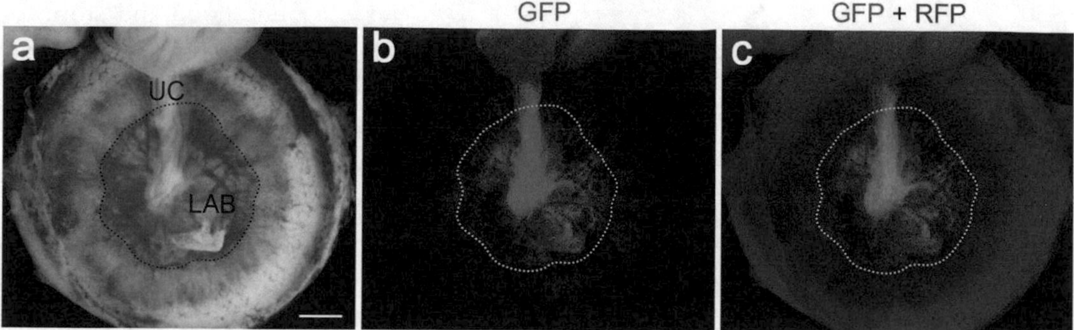

Fig. 1 E12.5 mouse placenta labyrinth. (**a**) Image of the embryonic side of a placenta still attached to an E12.5 embryo through the umbilical cord (UC). The labyrinth (LAB) is delimited by a dotted line. (**b**) GFP labels the labyrinth. (**c**) Fluorescence micrograph showing the labyrinth labeled by GFP (green). tdTomato RFP (red) labels nonendothelial cells. Scale bar = 1000 μm

3. Transfer the digested tissue into a 15 ml conical tube and add the same volume of TrypLE™ added in **step 1** of prewarmed DMEM containing 15% of FBS to stop the digestion.
4. Centrifuge the cell suspension at 350 × *g* for 5 min at 4 °C and discard the supernatant being careful not to disturb the pellet.
5. Resuspend the pellet with 10 volumes of 1× Red Blood Cell Lysis Solution at room temperature (*see* **Note 2**).
6. Centrifuge the cell suspension at 350 × *g* for 5 min at 4 °C, aspirate the supernatant completely (*see* **Note 11**).
7. Resuspend the cell pellet in 10 ml of 1× PBS, Centrifuge under the same conditions as above.
8. Gently resuspend the cell pellet into 500 μl of cold DMEM +1% FBS + 1 mM EDTA in a 5 ml polypropylene round-bottom tube.
9. Add 2 μl/ml of propidium iodide (PI) (*see* **Note 3**).
10. Pipette the cell suspension through a cell strainer into a 5 ml polypropylene round bottom tube (*see* **Note 12**).
11. For cell collection, transfer 1 ml of cold cell suspension into a 5 ml round bottom tube.
12. The cell suspension can now be loaded into a cell sorter. We have used the MoFlo Astrios cell sorter (*see* **Note 13**).
13. Establish the proper gates to distinguish GFP positive (endothelial) from GFP negative (nonendothelial) cells (Fig. 2).

3.3 Endothelial Cell Culture

1. Count the endothelial cells using a hemocytometer.
2. Seed 12,000 cells in 100 μl of ECM onto the 0.1% gelatin- or Matrigel-coated well in 96-well cell culture plates. Incubate the plate in a cell culture chamber at 37 °C in presence of 5% CO_2 (*see* **Note 14**).

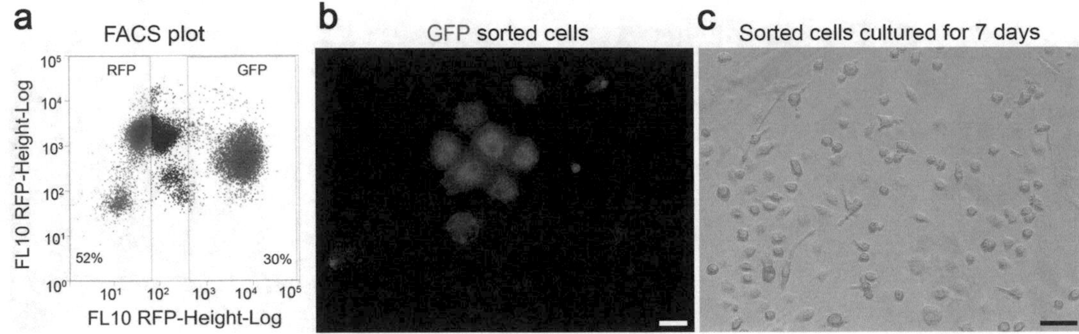

Fig. 2 FACS and culture of GFP positive placental endothelial cells. (**a**) The proportion of GFP+ and GFP- (RFP+) cells in FACS plots. (**b**) Fluorescence microscopy showing GFP+ cells. Scale bar = 25 μm. (**c**) Bright field image of sorted cells cultured for 7 days on 0.1% a gelatin-coated 96-well plate. Scale bar = 100 μm

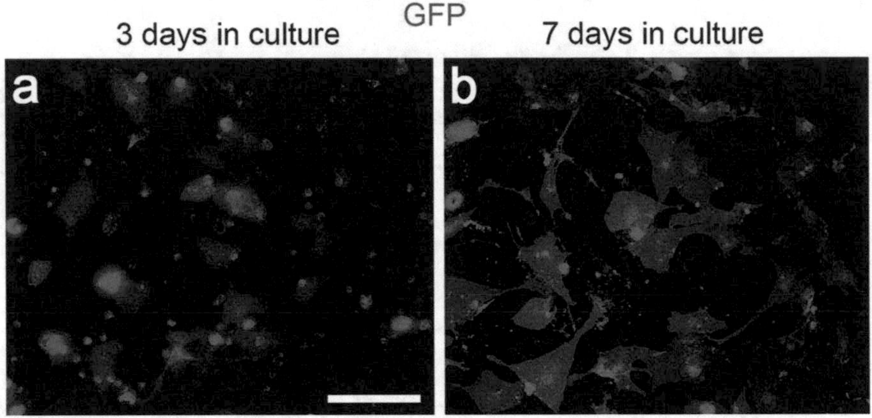

Fig. 3 Labyrinth endothelial cells in the culture evidenced by GFP fluorescence. (**a**) Labyrinth ECs cultured for 3 days. (**b**) Labyrinth ECs cultured for 7 days. Scale bar = 50 μm

3. Change the medium after incubating for 6 h (or overnight). Aspirate the medium and wash with prewarmed PBS to remove the dead cells.
4. Change the medium every 3 days afterward.
5. The placenta endothelial cells, which can be visualized under fluorescence microscopy for GFP, appear elongated after 7 days in culture (Fig. 3b). The cells reach 50–60% confluence after 7 days, and 70–80% confluence after 14 days in culture.

3.4 CD31 Staining on Labyrinth Endothelial Cells

The cells can be immunostained using an anti-CD31 antibody in combination with other antibodies to evidence other proteins of interest.

1. Fix the cells by adding 300 μl of 4% PFA per well in a 12-well plate, incubate for 5 min at room temperature.
2. Wash three times with PBS, with 5 min incubation at room temperature in between.

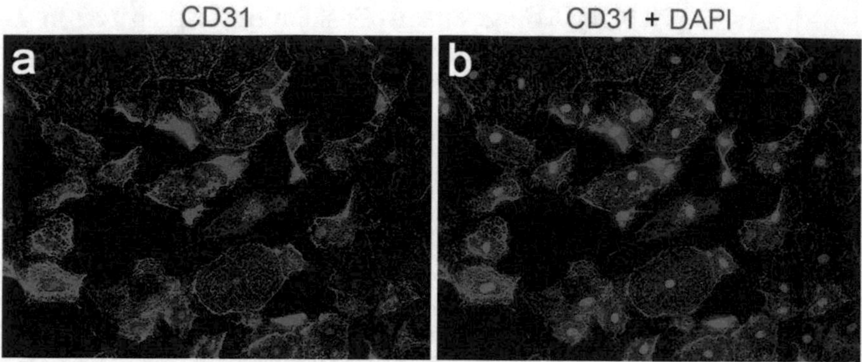

Fig. 4 CD31 staining on labyrinth endothelial cells cultured for 10 days on a 0.1% gelatin-coated well. (**a**) CD31 signal in green identifies ECs. (**b**) Merge with DAPI revealing nuclei. Cells were imaged using a Spinning Disk Confocal Microscope. Scale bar = 50 μm

3. Add 300 μl of blocking buffer to the cells and incubate for 30 min in a humidified chamber at room temperature.
4. Add primary antibodies (CD31 (1200)) in incubation buffer and incubate overnight at 4 °C in a humidified chamber to avoid drying of cells.
5. Wash the cells with 100 μl washing buffer three times with 5 min incubation at 4 °C in between.
6. Add secondary antibodies (Alexa Fluor 488 goat anti-rat IgG, or Alexa Fluor 546 goat anti-rabbit IgG (1:700), depending on the host of the anti-CD31 antibody) and DAPI (1:1000), diluted in incubation buffer, and incubate for 1 h at room temperature.
7. After washing with 1× PBS for three times for 5 min each, cover the cells with mounting medium and place a coverslip on top before imaging.
8. Cells can be observed and imaged under fluorescent light microscopy, or spinning disk confocal microscope using appropriate lasers (Fig. 4)

4 Notes

1. The day when a vaginal plug is first detected is considered as E0.5.
2. The red blood cell lysis solution (1×) is diluted with water at room temperature before use.
3. The Propidium Iodide (PI) stock solution is prepared at 1 mg/ml, and the concentration of the work solution is 2 μg/ml.

4. The 0.1% gelatin work solution is prepared in 1× PBS at pH 7.4, autoclaved and stored at 4 °C. Prewarm the solution in a water bath at 37 °C for 15 min before use.
5. Dilute the human vascular endothelial growth factor (hVEGF) in 0.1% FBS in 1× PBS pH 7.4. The stock solution is 50 μg/ml. Store the stock at −20 °C. Store for long term at −80 °C.
6. Thaw and keep the Matrigel matrix on ice at all times. Before pipetting the Matrigel, precool the pipet tips at 4 °C for 20 min. Pour enough Matrigel (60 μm/well in 96-well plate) to generate a thin layer (0.5 mm) to coat the bottom of the wells.
7. The human vascular endothelial growth factor (hVEGF) is added into the ECM always before Use.
8. DAPI should not be dissolved directly into PBS, as it will be degraded. The solutions are stable in the dark at 2–8 °C for several weeks.
9. Cut the labyrinth tissue into small pieces around 1×1×1 mm^3 by pulling with straight dissecting forceps. Rinse the tissue in cold 1× PBS twice, or more as required to remove blood.
10. The labyrinth endothelial cells are dissociated in TrypLE™ Express solution. The amount of TrypLE™ Express solution is adjusted to the amount of tissue. We use 1 ml of TrypLE™ Express for every two labyrinths.
11. Keep the cell suspension on ice and always centrifuge at 4 °C.
12. The labyrinth cell suspension must be pipetted through the cell strainer twice in order to obtain a uniform single cell suspension. This also helps to eliminate the bigger trophoblasts. First pass the cell suspension through a 100 μm pore size cell strainer, then through a 70 μm pore size cell strainer, into a 5 ml polypropylene round-bottom tube. The single cell suspension is ready to FACS after this.
13. The MoFlo Astrios is a Beckman Coulter sorter; the UV lasers used detect green, red, blue, and violet in the same sorting.
14. The 0.1% gelatin solution has to be prewarmed in a 37 °C water bath for 15 min before use. The Matrigel Matrix must be kept on ice. Coat the plates with 0.1% gelatin or Corning Matrigel Matrix, and make sure they cover the entire surface area. Place the plates in a cell culture incubator at 37 °C. After 1 h remove the gelatin solution and rinse the well with prewarmed PBS twice. The Matrigel will solidify. The plates are now ready for seeding the cells with the appropriate volume of ECM.

Acknowledgments

We thank The SickKids-UHN Flow Cytometry Facility for help with FACS, and The Centre for Phenogenomics (TCP) for mouse husbandry. This work was supported by the Heart and Stroke Foundation of Canada (G-17-0018613), the Natural Sciences and Engineering Research Council of Canada (NSERC) (500865), the Canadian Institutes of Health Research (CIHR) (PJT-149046), and Operational Funds from the Hospital for Sick Children to P.D.-O.

References

1. Larrivee B, Karsan A (2005) Isolation and culture of primary endothelial cells. Methods Mol Biol 290:315–329
2. Nachman RL, Jaffe EA (2004) Endothelial cell culture: beginnings of modern vascular biology. J Clin Invest 114(8):1037–1040. https://doi.org/10.1172/JCI23284
3. Gulino-Debrac D (2013) Mechanotransduction at the basis of endothelial barrier function. Tissue Barriers 1(2):e24180. https://doi.org/10.4161/tisb.24180
4. Benelli R, Albini A (1999) In vitro models of angiogenesis: the use of Matrigel. Int J Biol Markers 14(4):243–246
5. Cheung K, Ma L, Wang G, Coe D, Ferro R, Falasca M, Buckley CD, Mauro C, Marelli-Berg FM (2015) CD31 signals confer immune privilege to the vascular endothelium. Proc Natl Acad Sci U S A 112(43):E5815–E5824. https://doi.org/10.1073/pnas.1509627112
6. Dong QG, Bernasconi S, Lostaglio S, De Calmanovici RW, Martin-Padura I, Breviario F, Garlanda C, Ramponi S, Mantovani A, Vecchi A (1997) A general strategy for isolation of endothelial cells from murine tissues. Characterization of two endothelial cell lines from the murine lung and subcutaneous sponge implants. Arterioscler Thromb Vasc Biol 17(8):1599–1604
7. Marelli-Berg FM, Peek E, Lidington EA, Stauss HJ, Lechler RI (2000) Isolation of endothelial cells from murine tissue. J Immunol Methods 244(1–2):205–215
8. Jin GZ, Park JH, Lee EJ, Wall IB, Kim HW (2014) Utilizing PCL microcarriers for high-purity isolation of primary endothelial cells for tissue engineering. Tissue Eng Part C Methods 20(9):761–768. https://doi.org/10.1089/ten.TEC.2013.0348
9. Muzumdar MD, Tasic B, Miyamichi K, Li L, Luo L (2007) A global double-fluorescent Cre reporter mouse. Genesis 45(9):593–605. https://doi.org/10.1002/dvg.20335
10. Kisanuki YY, Hammer RE, Miyazaki J, Williams SC, Richardson JA, Yanagisawa M (2001) Tie2-Cre transgenic mice: a new model for endothelial cell-lineage analysis in vivo. Dev Biol 230(2):230–242. https://doi.org/10.1006/dbio.2000.0106

Chapter 11

Flow Cytometry and Lineage Tracing Study for Identification of Adipocyte Precursor Cell (APC) Populations

Ju Hee Lee, Azadeh Yeganeh, Hisato Konoeda, Joon Ho Moon, and Hoon-Ki Sung

Abstract

Flow cytometry and fluorescence-activated cell sorting (FACS) techniques have significantly advanced the characterization of adipocyte precursor cell (APC) populations. They allow immunophenotyping, quantification, and isolation of distinct populations, which is critical for understanding adipose tissue development and homeostasis. Here, we describe the identification and purification of adipocyte precursor cells using flow cytometry and FACS, defined by previously established surface marker profiles. In addition, we describe the mouse models and whole adipose tissue visualization techniques that will enable us to characterize the plasticity and the cellular origin of APCs.

Key words Flow cytometry, Adipocyte precursor cells (APCs), Lineage tracing, Cre-recombinase transgenic mouse model, Adipose tissue biology

1 Introduction

Adipose tissue is a loose connective tissue found ubiquitously throughout the body. It is traditionally classified into two types—white adipose tissue (WAT), which stores excess energy in the form of triglycerides, and brown adipose tissue (BAT), which generates non-shivering thermogenesis [1]. The current epidemic of obesity and associated metabolic diseases, such as type 2 diabetes, demands a better understanding of how adipocytes develop and where they originate.

Various adipose tissue depots are primarily distinguished by their anatomical locations. Two major types of WAT are dispersed throughout the body: visceral WAT, distributed around internal organs, and subcutaneous WAT, found underneath the skin. The WAT develops in utero but mainly occurs following birth and throughout the lifetime. In humans, the large depot of

Paul Delgado-Olguin (ed.), *Mouse Embryogenesis: Methods and Protocols*, Methods in Molecular Biology, vol. 1752,
https://doi.org/10.1007/978-1-4939-7714-7_11,

BAT is present in newborns, but only small amounts persist throughout adulthood. In rodents, BAT is mostly distributed in the interscapular and perirenal regions [1, 2]. Brown adipocytes express high level of a mitochondrial protein called uncoupling protein-1 (UCP1), owing to their requirement of high amounts of ATP to carry out non-shivering thermogenic processes [2]. Unlike WAT development, which majorly occurs in the postnatal period, BAT development occurs before birth [3]. Recently, a new type of brown-like adipocytes, called "beige/brite adipocytes," that shows distinct characteristics from WAT and BAT were identified. These brown-like adipocytes reside within WAT and are activated upon various environmental and physiological stimuli, such as cold exposure, β3-adrenergic activators, exercise, and fasting [4–6].

WAT and BAT develop from the mesoderm, but they are derived from distinct precursor populations. Mesenchymal stem cells are committed to Myf5-negative (WAT) or Myf5-positive (BAT and skeletal muscle) precursor lineages [7]. Several adipogenic transcriptional cascades, such as peroxisome proliferator-activated receptor-gamma (Pparγ) and CCAAT/enhancer-binding protein alpha (C/EBPα), regulate differentiation of WAT and BAT [8]. Recent research suggests that WAT can undergo browning, in terms of morphology, gene expression profile (i.e. UCP1 expression), and higher mitochondrial activity [9]. A potent inducer of beige adipocyte activation is cold stimulation [10, 11]. An in vivo lineage tracing study using UCP1-labeling in mice has demonstrated that acute cold exposure results in appearance of $UCP1^{+}$ beige adipocytes in subcutaneous WAT. When returned to ambient temperature, the beige adipocytes loose their UCP1 expression and revert to a WAT phenotype. Subsequent cold exposures result in rebrowning of former beige adipocytes, along with newly formed beige adipocytes [12]. This study suggests the bi-directional interconversion of white and beige adipocytes and the adaptive physiological response of WAT under environmental challenges. The ability of brown/beige adipocytes to suppress obesity through increased energy expenditure sparks an interest in the development and metabolic function of beige adipocytes and their therapeutic potential for obesity and diabetes.

Here, we describe experimental methods for the identification and purification of adipocyte precursor cells using flow cytometry, FACS, and a technique to visualize whole adipose tissue. In addition, we describe the mouse models that will enable us to characterize the plasticity and the cellular origin of APCs.

2 Materials

Prepare and store all reagents at 4°C unless indicated otherwise.

2.1 Cell Preparation for Flow Cytometry

1. Tissue digestion cocktail: 1 mg/mL collagenase II in DMEM/F12 media and mix well. To 500 mL of media, add 5 mL fatty acid-free BSA and 1 mL of 500 UG DNase I. Filter the cocktail through 0.2 μm filter. Make aliquots and store at −80°C until use.
2. FACS buffer: 1 × HBSS (without sodium bicarbonate and phenol red), 1 mM prefiltered EDTA, 25 mM HEPES solution, 1% FBS.
3. Wash buffer: 1 × PBS, 1% FBS.
4. Wash media: DMEM/F12 media, 1% FBS.
5. Sterile petri dishes and surgical tools.
6. Rotator.
7. Cell strainers: 40 μm, 200 μm.
8. Red blood cell lysis buffer.
9. CD16/CD32 antibody (Fc blocker).
10. Viability dye.
11. Antibodies of target markers labeled tagged with fluorophore-conjugated secondary antibodies.
12. Round bottom FACS test tubes with 35 μm-filtered cap.
13. Compensation beads.

2.2 Cell Sorting and Adipocyte Differentiation

1. Cell collecting media: DMEM/F12 media, 50% FBS. Aliquot in Eppendorf tubes and store until use.
2. Plating media: DMEM/F12 media, 15% FBS.
3. Adipocyte differentiation media: DMEM/F12 media, 1 μM indomethacin, 0.25 μM dexamethasone, 1 μM insulin, 0.5 mM isomethylbutylxanthine.
4. Multiwell cell culture plates.

2.3 Whole Mount Staining of Adipose Tissue and Microscopy

1. Fixation solution for whole-mount imaging: Prepare a fresh batch of 1% paraformaldehyde (PFA) in 1 × PBS before each use.
2. Blocking solution: Add 500 μL of goat serum plus 10 μL of Triton X-100 in 10 mL of 1 × PBS and mix well.
3. Mounting media with DAPI.

3 Methods

3.1 Using Cre/Lox System to Study Adipose Biology

The Cre-loxP recombination system is a tool used for site-specific manipulation of genes of interest, where Cre expression can be targeted at a cell/tissue-specific level. This system has revolutionized the study of the function and the development of adipose tissue in vivo. More recently, the development of in vivo Cre induction by administration of tamoxifen or doxycycline allows observation of spatiotemporal regulation of a target gene [13]. Several adipocyte-specific promoters have been generated to express Cre recombinase only in adipose tissues. Some of the commonly used adipose-specific Cre transgenes include aP2 (Fabp4), AdipoQ, Ng2, Pparγ, PdgfR-α, and PdgfR-β [14–16]. Various fluorescent reporters are combined with Cre transgenes to define Cre activity in a spatiotemporal manner (*see* below).

3.2 Using In Vivo Lineage Tracing to Track Adipocyte Precursor Cells

Lineage tracing technology has long been used to identify the progeny of a single cell, and providing a better understanding of developmental dynamics and the fate and plasticity of stem cells [17]. In combination with recent advances in cell-type specific Cre recombinase (Cre)-expressing models, multicolor reporter mouse lines (e.g., Brainbow and mT/mG mouse) have provided unprecedented insight into various scientific questions, including the origin of an adipocyte and tracking of adipocyte precursor populations [14, 16, 18, 19]. In this system, adipose-specific expression of Cre turns on the fluorescent reporter protein in a Cre expressing cell-specific manner, allowing visualization/tracing of the adipocyte progenitor and their descendant adipocytes.

For studies of white adipocytes (WAs), the traditional reporters that mainly stain the cytoplasm making it difficult to visualize adipocytes due to the lack of cytoplasm in mature WAs. To overcome this limitation, mT/mG (membrane-tdTomato/membrane-GFP), a reporter that marks Cre excision by switch from membrane-targeted tdTomato to membrane-targeted GFP, is a commonly used genetic tool in adipose biology. It is a double-fluorescent system that allows direct visualization of Cre-excised and non-excised cells at a single cell resolution (Fig. 1). The localization of fluorescent proteins to the membrane allows visualization of entire adipocyte morphology, making it an ideal reporter system for studying adipocytes [19, 20].

3.3 Characterization of Adipose Precursor Populations Using Flow Cytometry and Fluorescence-Activated Cell Sorting (FACS)

Adipose precursor cell populations have been well characterized in mice using flow cytometry and fluorescence-activated cell sorting (FACS) to isolate populations of interest based on a panel of cell surface marker expression. Using FACS, two white adipocyte precursor cell populations have been identified in mice. Both populations are negative for blood cell and endothelial-specific lineage

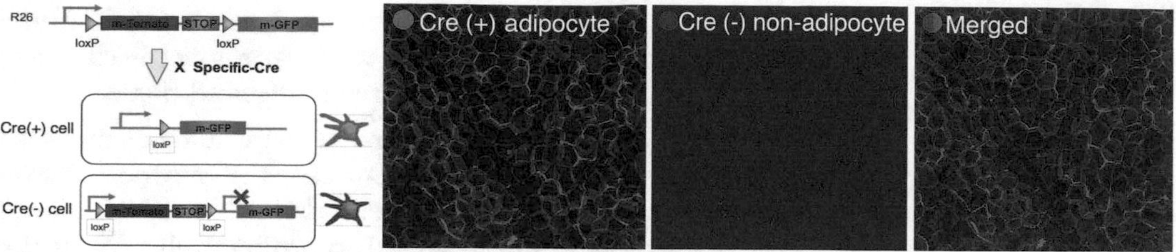

Fig. 1 Lineage tracing system for adipose tissue. Cre (+) cells express mGFP by excising two loxP sites while Cre (−) cells express mTomato

markers (CD45, Ter119, CD31), and positive for mesenchymal and stem cell markers (CD29, CD34, Sca1). These populations are separated by their expression of CD24 ($CD24^+$ or $CD24^-$). $CD24^-$ populations are characterized as preadipocytes as they exhibit limited adipogenic capacity when transplanted in lipodystrophic A-Zip mice. In contrast, $CD24^+$ cells are highly adipogenic in culture and induce adipogenesis when transplanted in A-Zip mice [19, 21]. Both $CD24^+$ and $CD24^-$ cell populations commonly express PdgfR-α, a well-known adipocyte lineage marker. Studies have also demonstrated the importance of angiogenesis in WAT development and expansion. Adipogenesis is tightly associated with and modulated by angiogenesis in a spatiotemporal manner throughout prenatal and adult life. Multiple studies have demonstrated the importance of adipocyte development in coordination with the vasculature [22–25]. It is suggested that perivascular mural cells (or pericytes) are a source of mesenchymal stem cells, and that a certain adipocyte population may be derived from a mural cell origin. For example, a study using Pparγ lineage tracing demonstrated that the vascular walls of adipose tissue blood vessels represent a niche for white adipocyte precursor cells. In particular, PdgfR-β marks mural cells, and the adipose PdgfR-β^+ stromal vascular (SV) cells display higher adipogenic potential than PdgfR-β^- SV cells [22].

3.4 Protocol for Identifying Adipose Progenitor Cells Using Flow Cytometry Analysis

3.4.1 Cell Preparation and Antibody Staining

1. Prepare tissue digestion cocktail as previously indicated (Cocktail solution should be stored as aliquots in −80°C).
2. Harvest mouse adipose tissue depots (inguinal, perigonadal, perirenal and mesenteric).
3. When dissection is complete, transfer the adipose tissue to a sterile petri dish on ice.
4. Mince the tissue to a fine consistency (1–2 mm^2) to form a paste using sterile scissors (*see* **Note 1**).
5. Add approximately 10 mL of the digestion cocktail into the petri dish with the finely minced tissue. Suspend the minced tissue in the enzyme cocktail and transfer to a 50 mL tube.

6. Incubate for ~30–45 min in a rotating shaker at 37°C. In the last 5 min of digestion, shake the tube vigorously (5–10 times) by hand to assist breaking down the undigested tissue.
7. After digestion is complete, add twice the volume of the wash medium to the tissue digest to neutralize the enzyme and stop digestion. Filter the tube content through a 200 μm cell strainer into a new sterile 50 mL centrifuge tube. Wash the filter with 5 mL of the wash buffer. Discard undigested tissue on the strainer. (*see* **Note 2**)
8. Centrifuge the tube content at 600 × *g* for 5 min to sediment cells.
9. Pipette out the floating top fraction (mature adipocytes) and remove the supernatant from the tube without disturbing the bottom-layer, reddish-appearing stromal vascular fraction (SVF) pellet.
10. Resuspend the pellet in 2 mL wash buffer. Wash and separate the cells thoroughly by gently pipetting. Then filter the mixture through a 40 μm cell strainer into a new sterile 50 mL centrifuge tube. Wash the filter with 5 mL of wash buffer.
11. Centrifuge the tube content 600 × *g* for 5 min to sediment cells.
12. Discard the supernatant and resuspend the pellet in 2–3 mL of red blood cell lysis buffer.
13. Incubate at room temperature for 5 min, and then add an equal volume of the wash buffer.
14. Centrifuge at 600 × *g* for 5 min and discard the supernatant. If the cells still appear red, repeat **steps 12–14**.
15. Resuspend the cells in ~200–500 μL of FACS buffer. (*see* **Note 3**)
16. Add 2 μL of Fc-block (CD16/CD32) per ~200 μL of cells and incubate on ice for 10–15 min.

    Before proceeding to the following step, prepare an antibody staining cocktail containing all fluorochromes of interest, with a viability dye (concentration of 1:500–1:1000), for all experimental samples. In addition to the experimental samples, four sets of control samples are required. (1) FMO (fluorescence-minus-one) samples, including every antibody except one fluorochrome. For example, FITC FMO sample will contain every other antibody except the FITC antibody. (2) Single antibody staining, prepared with suitable compensation beads containing each antibody. (3) Unstained sample, prepare one tube containing unstained cells. (4) Viability control, a tube containing cells stained with viability dye such as DAPI (4′,6-diamidino-2-phenylindole), a florescence dye that binds to A-T rich regions in DNA and marks the dead cells (or any other suitable viability dye) (*see* **Note 4**)

- A common adipose precursor cell population panel:
- CD31$^-$, CD45$^-$, CD29$^+$, CD34$^+$, Sca1$^+$, CD24$^+$ (for adipocyte progenitor population) or CD24$^-$ (for preadipocyte population).
- When using a mouseline with a fluorescent reporter, such as mT/mG mice, avoid using fluorochromes that overlap with the fluorescent reporter of the mouse.

17. Add cells suspended in FACS buffer into each tube with prepared antibody cocktail.
18. Incubate the tubes for 30–40 min at 4°C, protected from light.
19. Wash stained cells with 300–500 μL of FACS buffer and centrifuge for 10 min at 600 × *g*, then remove the supernatant. Repeat this step.
20. Resuspend the cells in final volume of 300–500 μL of FACS buffer, then pipette the cells through the 70 μm-filtered FACS tubes. Place on ice, protected from light, until loading in the flow cytometry machine.

3.4.2 Flow Cytometry Analysis

Using a flow cytometer such as GALLIOS (Beckman) with 4 lasers and 10 detectors enables us to detect 10 different fluorochromes simultaneously.

1. Read the unstained cells as a negative control to eliminate the background signal on all the channels.
2. Next read the unstained cells stained only with DAPI or other viability marker.
3. Then, read the compensation beads stained with single antibody to find the spectral overlap among the fluorochromes. These differences could be removed by changing the voltages or by compensation during the analysis.
4. Read the experimental tube stained with all the antibodies followed by the FMO control tubes. (In multicolor flow cytometry, the FMO controls will be used for gating different cell populations during analysis.)
5. To analyse the flow cytometry data, select the correct cell population to analyse. For two step-gating, use Forward Scatter (FS) and Side Scatter (SS) to discriminate the cells based on size and granularity, respectively. Doublets and aggregates have higher time of flight (TOF) and they could be eliminated at this step.
6. Since dead cells bind non-specifically to many reagents and antibodies, they could lead to false signals in flow cytometry. Thus, using the data from tube containing cells solely with

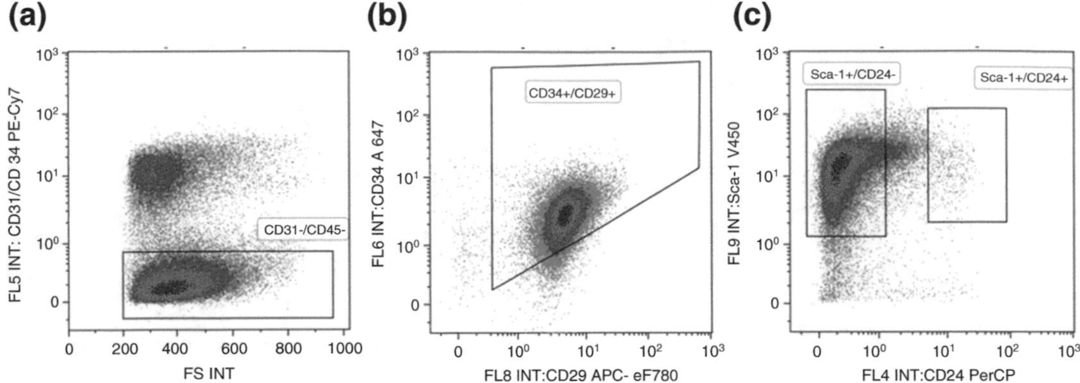

Fig. 2 Flow cytometry plots identifying adipocytes progenitor cells. (**a**) CD31⁻ and CD45⁻ population, using antibodies conjugated with PE-Cy7 fluorochrome. (**b**) CD34⁺ and CD29⁺ population, using antibodies conjugated with Alexa Fluor 647 and APC-eFluor 780 fluorochrome respectively. (**c**) Sca-1⁺ and CD24⁺ and CD24⁻ populations, using antibodies conjugated with V450 and PerCP fluorochrome, respectively

viability marker, dead and/or apoptotic cells can be eliminated. Use FS and dead-cell stain parameters to gate the live cells.

7. Now that viable populations are selected, gate the cells using the FMO samples. As the Rodeheffer group [21] demonstrated, gate the lineage-negative cells first, which are defined as $CD31^-$ and $CD45^-$. Next, the selected population will be gated for $CD29^+$ and $CD34^+$. Subsequently, the double-positive population will be gated for $Sca1^+$ and $CD24^+$ or $CD24^-$ population based on population of interest (Fig. 2).

3.4.3 Fluorescence-Activated Cell Sorting (FACS)

FACS enables the isolation of unique cell populations based on cell surface profile, and it serves as a useful tool in purifying adipose progenitor cells.

The general protocol for FACS is similar to flow cytometry. However, the procedure has to be carried out in sterile environment using sterile reagents. Similar control tubes are prepared for FACS. Depending on the number of desired cell population, sterile eppendorf tubes containing collecting medium are prepared. Examples of FACS plot shown in Fig. 3.

1. After FACS, centrifuge cells at 300 × *g* for 5 min and resuspended in 100–500 μL plating medium depending on the number of sorted cells. Approximately 10,000–20,000 cells are seeded in each well in a 96-well plate (*see* **Note 5**).
2. Swirl cells for even distribution and incubate at 37°C. The seeded cells should not be disturbed for 24–48 h.
3. Replace media after 48 h to DMEM/F12 containing 10% FBS.

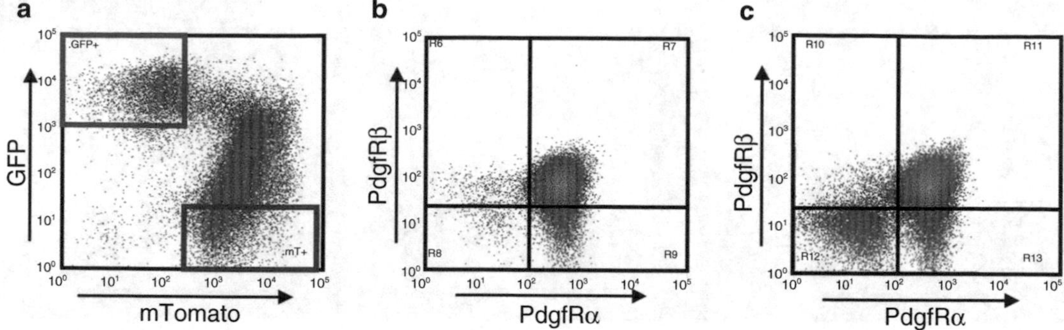

Fig. 3 FACS plots from Ng2 Cre; mT/mG mice. (**a**) SVF analyzed by GFP (mG) and Tomato (mT). (**b**) mG population analyzed by PdgfR-α and PdgfR-β using PE-Cy7 and APC fluorochrome, respectively (**c**) mT population analyzed by PdgfR-α and PdgfR-β using PE-Cy7 and APC fluorochrome, respectively

4. The confluent cells can be differentiated to adipocytes using the adipogenic differentiation media. After 2 days, media should be changed to DMEM/F12 containing 10% FBS with Insulin (1 μM). Lipid droplet filled adipocytes will be observed after 4 days of differentiation.

3.5 Whole Mount Staining of Adipose Tissue and Visualization

Adipose tissue can be observed under the microscope without sectioning, using a technique called whole mount staining. This method is mostly used for adipose depots with less cells and lower vessel density, such as perigonadal adipose tissue (Fig. 3). To prepare for whole mount staining of adipose tissue:

1. Dissect the adipose tissue with blunt-ended tools to avoid stretching or damaging the tissue.
2. Fix the tissue in fixation solution overnight at 4°C or 1–2 h at RT.
3. Wash the tissue on ice with cold PBS for 10 min, using a shaker. Repeat this step three times.
4. If using a multicolor reporter mouse line such as mT/mG mouse, the tissue is ready for microscopy. However, if further antibody staining is required, follow the next steps.
5. Permeabilize and block the tissue using the blocking solution for 1 h.
6. Incubate with primary antibodies, diluted in blocking solution, overnight at 4°C.
7. On the next day, wash the tissue on ice using cold PBS for 10 min, using a shaker. Repeat this step three times.
8. Incubate with fluorescence-conjugated secondary antibodies diluted in PBS with 1% BSA for 1 h at room temperature.
9. Wash the tissue as explained on **step 3**.

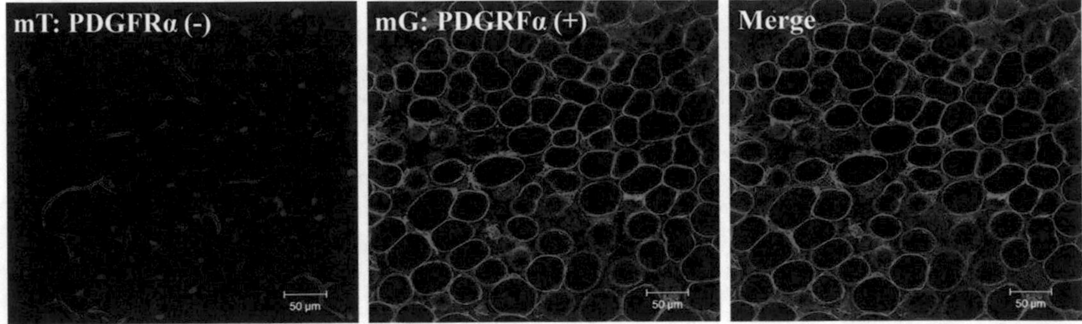

Fig. 4 Whole mount staining of perigonadal adipose tissue of 8 weeks old, PdgfR-α Cre; mT/mG mouse. While PdgfR-α derived cells (mG) shown in green, cells that have not expressed PdgfR-α during their life span shown in red (mT)

10. The tissue is ready for microscopy. Apply nonsolidifying mounting media and visualize the tissue using confocal microscope. If nuclear staining is required, use mounting media containing DAPI. An example of adipose tissue whole mount staining shown on Fig. 4.

4 Notes

1. Avoid over-digestion (>3 min of chopping) of tissue as it can increase cell death. Dissect out the lymph node from inguinal or mesenteric adipose depot if needed.
2. Before washing the unfiltered tissue, gently mash the undigested tissue pieces with a blunt tool and wash with extra washing buffer to increase cell yield.
3. Dilute 1 million cells in ~300 μL of FACS buffer. If cell number is unknown, dilute the cells in a lower volume of FACS buffer, and increase the volume accordingly.
4. When preparing cells for antibody staining, pool excess cells from samples with larger cell pellets to prepare for FMO samples.
5. To minimize cell death, it is ideal to seed the sorted cells as soon as possible.

Acknowledgments

This work was funded by grants from the Natural Sciences and Engineering Research Council (NSERC) of Canada, Pilot and Feasibility Study Grant of Banting & Best Diabetes Centre (BBDC), Centre for Healthy Active Kids (CHAK) Micro-grant and Sickkids Start-up fund to H.-K.S. J.-H.M is supported by the Restracomp fellowship from The Hospital for Sick Children.

References

1. Park A, Kim WK, Bae KH (2014) Distinction of white, beige and brown adipocytes derived from mesenchymal stem cells. World J Stem Cells 6(1):33–42
2. Harms M, Seale P (2013) Brown and beige fat: development, function and therapeutic potential. Nat Med 19(10):1252–1263
3. Moulin K et al (2001) Emergence during development of the white-adipocyte cell phenotype is independent of the brown-adipocyte cell phenotype. Biochem J 356(Pt 2):659–664
4. Uhm M, Saltiel AR (2015) White, brown, and beige; type 2 immunity gets hot. Immunity 42(1):15–17
5. Bostrom P et al (2012) A PGC1-alpha-dependent myokine that drives brown-fat-like development of white fat and thermogenesis. Nature 481(7382):463–468
6. Xu X et al (2011) Exercise ameliorates high-fat diet-induced metabolic and vascular dysfunction, and increases adipocyte progenitor cell population in brown adipose tissue. Am J Physiol Regul Integr Comp Physiol 300(5): R1115–R1125
7. Rosenwald M, Wolfrum C (2014) The origin and definition of brite versus white and classical brown adipocytes. Adipocytes 3(1):4–9
8. Rosen ED, MacDougald OA (2006) Adipocyte differentiation from the inside out. Nat Rev Mol Cell Biol 7(12):885–896
9. Wu J et al (2012) Beige adipocytes are a distinct type of thermogenic fat cell in mouse and human. Cell 150(2):366–376
10. Hui X et al (2015) Adiponectin enhances cold-induced Browning of subcutaneous adipose tissue via promoting M2 macrophage proliferation. Cell Metab 22(2):279–290
11. Odegaard JI et al (2016) Perinatal licensing of thermogenesis by IL-33 and ST2. Cell 166(4):841–854
12. Rosenwald M et al (2013) Bi-directional interconversion of brite and white adipocytes. Nat Cell Biol 15(6):659–667
13. Feil S, Valtcheva N, Feil R (2009) Inducible Cre mice. Methods Mol Biol 530:343–363
14. Jeffery E et al (2014) Characterization of Cre recombinase models for the study of adipose tissue. Adipocytes 3(3):206–211
15. Krueger KC et al (2014) Characterization of Cre recombinase activity for in vivo targeting of adipocyte precursor cells. Stem Cell Reports 3(6):1147–1158
16. Berry DC, Jiang Y, Graff JM (2016) Mouse strains to study cold-inducible beige progenitors and beige adipocyte formation and function. Nat Commun 7:10184
17. Kretzschmar K, Watt FM (2012) Lineage tracing. Cell 148(1–2):33–45
18. Livet J et al (2007) Transgenic strategies for combinatorial expression of fluorescent proteins in the nervous system. Nature 450 (7166):56–62
19. Berry R, Rodeheffer MS (2013) Characterization of the adipocyte cellular lineage in vivo. Nat Cell Biol 15(3):302–308
20. Muzumdar MD et al (2007) A global double-fluorescent Cre reporter mouse. Genesis 45(9):593–605
21. Rodeheffer MS, Birsoy K, Friedman JM (2008) Identification of white adipocyte progenitor cells in vivo. Cell 135(2):240–249
22. Tang W et al (2008) White fat progenitor cells reside in the adipose vasculature. Science 322(5901):583–586
23. Cao Y (2010) Adipose tissue angiogenesis as a therapeutic target for obesity and metabolic diseases. Nat Rev Drug Discov 9(2):107–115
24. Tran KV et al (2012) The vascular endothelium of the adipose tissue gives rise to both white and brown fat cells. Cell Metab 15(2): 222–229
25. Gupta RK et al (2012) Zfp423 expression identifies committed preadipocytes and localizes to adipose endothelial and perivascular cells. Cell Metab 15(2):230–239

Chapter 12

Whole-Mount and Section In Situ Hybridization in Mouse Embryos for Detecting mRNA Expression and Localization

Kazuko Koshiba-Takeuchi

Abstract

In situ hybridization is defined as one of the most useful and powerful methods to know where genes (e.g., mRNA, ncRNA) of interest are expressed in tissues. Expression of mRNA can be detected as blue or dark purple signals though hybridization, immunoreaction and coloring steps. Genome-wide approaches in various model animals have been conducted thoroughly, and have led to new research areas aimed at uncovering novel gene functions in cell differentiation and development. To elucidate gene function, spatiotemporal gene expression analysis is very important. Here I describe protocols of whole-mount and section in situ hybridization, and emphasize the relevance of optimizing temperature, and sodium concentration, in hybridization buffer and substrate to improve signal.

Key words In situ hybridization, Whole-mount, Section, Mouse embryo, mRNA, Expression analysis

1 Introduction

In situ hybridization is a method used to reveal the expression pattern of mRNA in whole embryos or in tissue sections. Originally, the hybridization probe was labeled with a radioisotope, and a photosensitive emulsion was used for the detection. A disadvantage of this method is that it could take a couple of weeks before getting to know the expression patterns of genes of interest [1]. In the 1990s, a nonradioisotope method was established, and the detection time was greatly shortened [2, 3]. Using a nonradioisotope probe labeled with digoxigenin (DIG), the expression of mRNA can be detected within 2–4 days. The antisense DIG-labeled RNA probe is synthesized from linearized plasmid containing the gene of interest. The antisense DIG-labeled probe is hybridized to endogenous mRNA in situ. An alkaline phosphatase (AP)-coupled anti-DIG antibody is then used to recognize the DIG-labeled probe. After antibody treatment, nitro blue tetrazolium (NBT)/5-bromo-4-chloro-3-indolyl phosphate (BCIP), or BM purple, are used as substrate for the AP to develop a blue, or dark purple signal, respectively (Fig. 1).

Paul Delgado-Olguin (ed.), *Mouse Embryogenesis: Methods and Protocols*, Methods in Molecular Biology, vol. 1752, https://doi.org/10.1007/978-1-4939-7714-7_12,

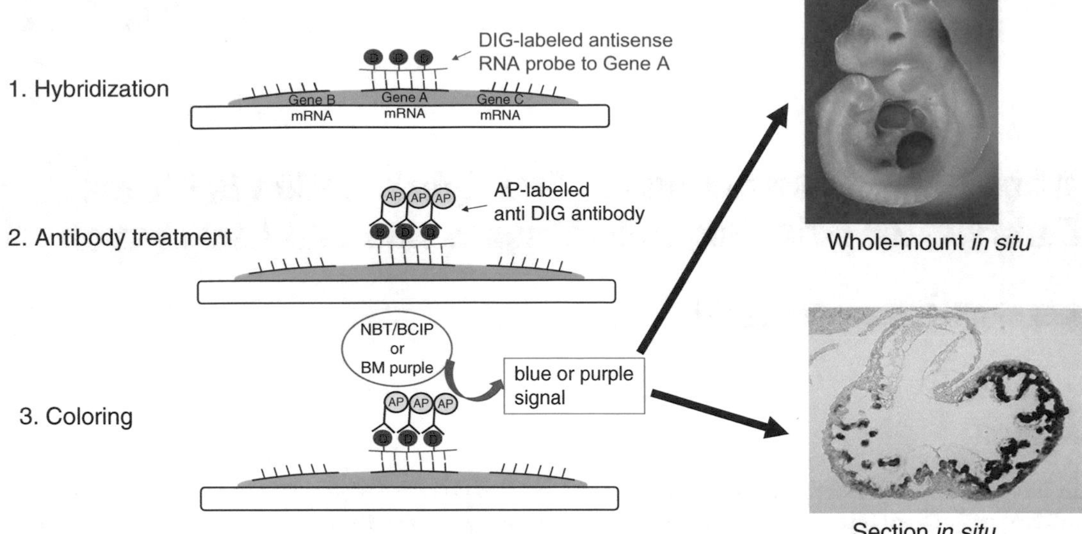

Fig. 1 Schematic of in situ hybridization

It is possible to observe the expression patterns of two different genes simultaneously by using a combination of Fluorescein-labeled RNA probe and DIG-labeled probe (two color in situ hybridization) [4]. However, in this chapter the basic protocol for in situ hybridization is described.

In southern and northern blotting, hybridization temperature or sodium concentration of the hybridization buffer and washing solution is modified depending on the probe (e.g., length, GC content, and target gene dosage) to improve specificity and intensity of the signal. Generally, a high stringency condition means high temperature and a low sodium concentration, suitable for long, or GC-rich probes. Conversely, a low stringency condition means a low temperature and a high sodium concentration, which facilitates hybridization of short, or low GC-content, probes. Stringency conditions determine the success of in situ hybridization. To optimize the hybridization condition for *Bmp10*, which is expressed in the trabecular layer of the heart, hybridization temperature, sodium concentration, and the substrate for alkaline phosphatase were tested in mouse heart sections. Here I show that different conditions of these factors affect the results of in situ hybridization on tissue sections.

2 Materials

2.1 Preparation of RNA Probe

The gene of interest cloned in a dual promoter vector (*see* **Note 1**) is used for RNA probe synthesis (Fig. 2a). For making the RNA probe, it is fine if the template gene is not full-length. A gene fragment

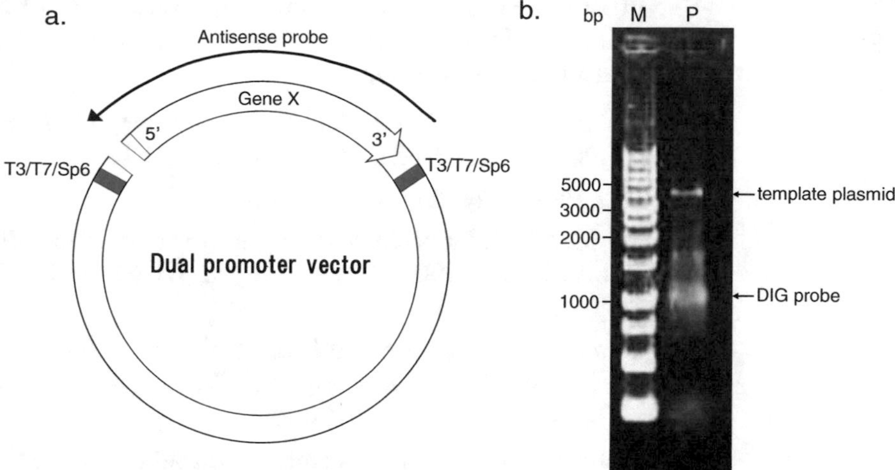

Fig. 2 Probe preparation. Image of a dual promoter vector (**a**) and detection of synthesized DIG probe on a 1% agarose gel (**b**). Left lane is the 1 kb marker (M). 1 μL of the reaction mix (P) was loaded in the right lane. Note that the linearized template plasmid and the probe are detected

corresponding to a unique, not conserved region can be used as template to reduce cross reaction with other related genes, and thus to increase specificity (*see* **Note 2**).

1. Treat the plasmid with a restriction enzyme at the 5′ end of the inserted gene to linearize the plasmid, and purify the linearized plasmid by using a PCR purification column (*see* **Note 3**).
2. Mix the followings; 2 μL 10× transcription buffer, 2 μL DIG-labeling mix, 1 μg linearized plasmid, 0.5 μL RNase inhibitor, 1 μL T3, T7, or Sp6 RNA polymerase, and bring the volume to 20 μL with DEPC-treated water (*see* **Note 4**). Use a RNA polymerase specific for the promoter positioned at the 3′ site of the cloned gene. Incubate the mixture for 2 h at 37 °C.
3. Check the transcription product on a gel by running 1 μL of the reaction (Fig. 2b).
4. When a clear band is observed, add 1 μL DNase (RNase-free), and incubate for 15 min at 37 °C
5. Add 90 μL of DEPC-treated water, 10 μL LiCl, and 300 μL EtOH to the reaction, mix well and leave at −20 °C for at least 1 h.
6. Spin for 10 min at 4 °C and remove the supernatant.
7. Wash with 500 μL 70% EtOH, spin for 5 min. After removing supernatant, air dry for 30 min.
8. Resuspend the pellet in 10 μL of DEPC-treated water and 90 μL hybridization buffer. Store at −20 °C. Before using, heat at 95 °C for 5 min to denature the probe.

2.2 In Situ Hybridization

For in situ hybridization, all solutions, except for formamide-containing solutions, should be filtered to remove dust, which may stain and adhere to the sample.

1. 20× PBS: 160 g NaCl, 4 g KCl, 23 g Na_2HPO_4, and 4 g KH_2PO_4 are dissolved in water and up to 1 L.
2. PBT: PBS containing 0.1% Tween 20.
3. 20× SSC: 175.3 g NaCl, and 88.2 g trisodium citrate dihydrate are dissolved in 800 mL water, adjust pH 4.5 with 1 M citric acid, and up to 1 L.
4. Hybridization solution: 50% formamide (deionized), 5× SSC pH 4.5, 50 μL/mL yeast tRNA, 1% SDS (filtrated), 50 μL/mL heparin, DEPC-treated water.
5. Wash solution 1: 50% formamide, 5× SSC pH 4.5, 1% SDS.
6. Wash solution 2 (washing solution for section in situ hybridization): 50% formamide, 2× SSC pH 4.5.
7. 10× TBS: 40 g NaCl, 1 g KCl, 125 mL 1 M Tris–HCl pH 7.5 are dissolved in water and up to 500 mL and autoclaved.
8. TBST: TBS containing 0.1% Tween 20.
9. Blocking solution: 10% heat-inactivated (56 °C, 30 min) sheep serum in TBST.
10. NTMT: 100 mM NaCl, 100 mM Tris–HCl pH 9.5, 50 mM $MgCl_2$, 0.1% Tween 20, and 2 mM levamisole.
11. Chamber solution: 50% formamide in 5× SSC pH 4.5.
12. Coloring solution for section in situ hybridization; 52.5 μL 100 mg/mL NBT, 53 μL 50 mg/mL BCIP in 150 mL NTMT.
13. Glass vial or netwells for whole-mount in situ hybridization.
14. Staining containers and a slide holder for section in situ hybridization.
15. Mounting reagent: Dissolve 2.4 g Mowiol in 6 mL glycerol and 12 mL 0.2 M Tris–HCl pH 8.5, mix well, and add water up to 24 mL. Incubate at 50 °C overnight. Spin down to remove undissolved Mowiol. Make aliquots and store at −20 °C.

3 Methods

3.1 Whole-Mount In Situ Hybridization

1. Dissect mouse embryos in PBS and remove yolk sac membranes. For embryonic day 10.5 (E10.5) and older embryos, remove the pericardium to increase penetration of fixative and probe (*see* **Note 5**).
2. Fix the embryos in 4% paraformaldehyde (PFA) in PBS at 4 °C overnight (*see* **Note 6**).

3. Wash in PBT 2× 5 min at 4 °C, shaking gently (*see* **Note 7**).
4. Dehydrate in a MeOH/PBT series. First, rinse in 25% MeOH in PBT for 10 min at room temperature (RT), gently shaking. Next rinse in 50% MeOH in PBT for 10 min, and in 75% MeOH in PBT for 10 min. Then wash in 100% MeOH 2× 10 min at RT (*see* **Note 8**).
5. Rehydrate embryos in 75%, 50%, 25% MeOH in PBT for 10 min each, and wash in PBT 2× 5 min at RT, gently shaking. Netwells can be used.
6. Bleach with 6% hydrogen peroxide in PBT 1 h at RT, gently shaking.
7. Wash in PBT 3× 5 min PBT at RT.
8. Treat with 10–20 μg/mL proteinase K in PBT for 15–30 min at RT. Generally E10.5 embryos are treated with 10 μg/mL proteinase K for 15 min (*see* **Note 9**).
9. Treat with 2 mg/mL glycine in PBT for 15 min at RT shaking (*see* **Note 10**).
10. Wash in PBT 2× 5 min at RT.
11. Post-fix with 4% PFA/0.2% glutaraldehyde in PBT for 20 min at RT.
12. Wash in PBT 2× 5 min at RT.
13. Prehybridize for 1 h at 70 °C in preheated hybridization buffer with a minimum volume (2–4 mL) (*see* **Note 11**).
14. Add a denatured probe directly to hybridization buffer in a 1–5 μL/100 μL dilution. Hybridize overnight at 70 °C (*see* **Note 12**).
15. Collect the hybridization solution and keep at −20 °C. It can be reused several times, and often works better than fresh solution (*see* **Note 13**).
16. Wash in washing solution 1, 3× 30 min at 70 °C gently shaking.
17. Wash in washing solution 2, 3× 30 min at 65 °C (*see* **Note 14**).
18. Wash in 50% washing solution 2 in TBST for 10 min at 65 °C, and cool down to RT.
19. Wash in TBST 3× 5 min at RT, gently shaking.
20. Treat embryos with blocking solution for at least 1 h at RT.
21. Remove blocking solution, add antibody solution (Dilute 1:2000–2500 anti-DIG antibody in 1% sheep serum in TBST) and incubate overnight at 4 °C, rocking gently.
22. Collect antibody solution. Keep it at 4 °C and reuse several times within a couple of months.
23. Wash in TBST 3× 5 min at RT shaking.

24. Wash in TBST 6-8× 30 min at RT, shaking (*see* **Note 15**).
25. Wash in NTMT 3× 10 min at RT, shaking.
26. Incubate with BM purple, cover with foil to avoid exposure to light and place on rocker. Check signal under a stereomicroscope after 15 min, and then every 30 min.
27. When a specific signal is detected, remove the BM purple.
28. Rinse 2× 5 min in NTMT and replace in PBT.
29. Post-fix with 4% PFA for 1 h at RT.
30. Wash in PBT 2× 5 min. Embryos are now ready for imaging under the microscope. The sample can be stored at 4 °C (*see* **Note 16**).

3.2 Section In Situ Hybridization

1. Prepare 8–10 μm paraffin sections or cryosections.
2. For paraffin sections, remove paraffin with Xylene for 10 min, and then dip into fresh Xylene for 10 min. Wash with 100% EtOH for 5 min, 90% EtOH for 5 min and 70% EtOH for 5 min.
3. Deparaffinized sections, or cryosections are washed in PBT for 10 min at RT.
4. Treat with 1 μg/mL Proteinase K in PBT at 37 °C for 7 min.
5. Wash in PBT 3× 5 min at RT.
6. Refix with 4% PFA for 20 min at RT.
7. Wash in PBT 2× 5 min.
8. Put 100 μL hybridization solution containing 1–5 μL probe on the sections on the glass slides, cover with a Parafilm and set slides on a wet box moisten with chamber solution (Fig. 3). Hybridize overnight at 65 °C (*see* **Note 17**).
9. Remove the Parafilm in heated washing solution and wash with a fresh washing solution 2× 30 min at 65 °C (*see* **Note 18**).
10. After washing with 50% washing solution in TBST for 10 min at 65 °C, cool down to RT.
11. Wash in TBST 3× 5 min.
12. Treat with 10% sheep serum in TBST (blocking solution) for 60 min at RT (*see* **Note 19**).
13. Put 100 μL antibody solution (1:1000 in 1% sheep serum in TBST) on slides, cover with Parafilm and set on wet box moistened with water. Leave it for 1 h at RT or 4 °C overnight.
14. Remove the Parafilm in TBST and wash in TBST 3× 5 min.
15. Rinse in NTMT for 5 min.
16. Prepare coloring solution in staining container, set slides and cover with foil. Leave overnight at RT.

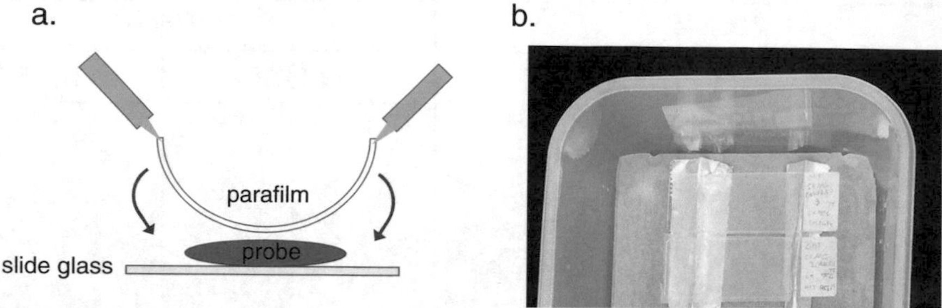

Fig. 3 Section in situ hybridization. To prevent bubble formation, pick up both ends of the Parafilm with forceps and extend from the middle toward the edges of the glass slide (**a**). Clean absorbent paper moistened with chamber solution is placed at the bottom of the container, which is covered with a lid. Folded foil bars can be used to maintain the slides separated from the bottom of the container (**b**). Seal the hybridization chamber tightly and hybridize for overnight

17. Check signal, and when specific signal without too much background is detected, stop the coloring. If the signals are still weak and the coloring solution turns purple, refresh the solution and leave it for one more night (*see* **Note 20**).
18. To stop the reaction, rinse in PBT for 10 min, fix with 4% PFA for 30 min at RT.
19. Rinse in PBT and mount with mounting reagent.
20. Testing different conditions show that the hybridization temperature of 70 °C is too high for section in situ hybridization, and it results in a weaker signal than 65 °C. In contrast, higher sodium concentration (6× SSC) slightly increases the signal. BM purple provides a clearer signal than NBT/BCIP (Fig. 4). When a weak signal is observed in the first trial, it is worth to try to use 6× SSC hybridization buffer and detect with BM purple.

4 Notes

1. Dual promoter vectors (e.g., pBluescript, pCRII) contain two promoters out of T3, T7, and Sp6, one at each end of the cloned fragment. The polymerase used for synthesis of the antisense probe must transcribe from the promoter at the 3′ end of the gene fragment.
2. A probe shorter than 200 bp might increase the possibility of unspecific hybridization and thus nonspecific signal. Probes at least 500 bp length are recommended.
3. For plasmid linearization, use a restriction enzyme leaving 5′-protruding, or a blunt end. A 3′-protruding end in the template would cause bidirectional transcription.

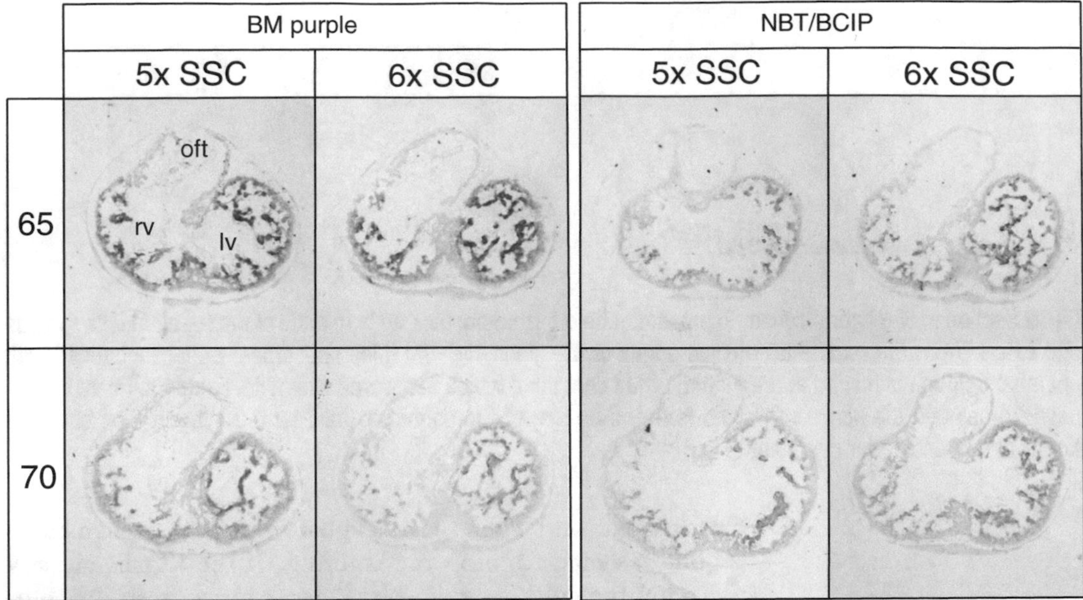

Fig. 4 Expression of *Bmp10* in mouse embryonic heart at E11.5 under various hybridization conditions. Hybridization temperature at 70 °C gives weak signal, while higher sodium concentration (6× SSC) gives a slightly stronger signal. BM purple can provide an intensified signal. rv: right ventricle, lv: left ventricle, oft: outflow tract

4. Preparing half volume of the reaction is enough if you are only testing probe quality.
5. Removal of extraembryonic membranes can be done after fixation, but it is recommended to remove before dehydration. After dehydration, the membrane sticks tightly to embryo.
6. Improper fixation, i.e., fixation for a short time or use of old PFA, might result in increased background. Fresh 4% PFA should be used.
7. After removing the extraembryonic membranes, make holes in the forebrain, midbrain, and hindbrain with dissection forceps to avoid probe trapping.
8. For older embryos, longer treatment time is better. For example, 30 min each for E14.5 and older, depending on the sample size. It is possible to store embryos in 100% MeOH at −20 °C for several months.
9. This step is important to reduce background. The concentration of proteinase K and treatment time should be adjusted depending on the size of embryos.
10. Dilute the 100× stock, which can be stored at −20 °C. Thaw in a 37 °C water bath before using.

11. To avoid dilution of hybridization buffer, you can add a rinse with a small volume of hybridization buffer before prehybridization.
12. First try a 1 μL/100 μL dilution, and if the signal is weak, increase the probe concentration.
13. In hybridization buffer, embryos become transparent; thus, extra caution should be payed not to suck the embryo when collecting hybridization buffer. Reused hybridization solution must be heat at 70 °C for 30 min before adding it to the prehybridized embryos.
14. When using netwells in this step, add Tween 20 at 0.1% to prevent embryos sticking to the net.
15. If there is not enough time to do coloring reaction during the day, it is fine to leave the sample in TBST at 4 °C overnight.
16. If a strong contrast is preferred, dehydrate embryos in a MeOH series. After dehydration, the embryo body turns white.
17. When covering with Parafilm, pay attention not to form any air bubble. Under high temperature, the air bubble will grow and prevent hybridization.
18. When removing the Parafilm, do not peel it off. Wait until the Parafilm floats away.
19. Keep the blocking solution at −20 °C and reuse.
20. When using BM purple, put 100 μL BM purple on the slide, cover with a transparent hybridization bag cut slightly wider than the slide. The signal should appear within several hours, so that check the signal every 30 min.

Acknowledgments

I thank F. Nakayama and D. Matsumura for technical assistance, and J. K. Takehchi for critical reading of the manuscript and useful advice.

References

1. Noji S, Yamaai T, Koyama E, Nohno T, Fujimoto W, Arata J, Taniguchi S (1989) Expression of retinoic acid receptor genes in keratinizing front of skin. FEBS Lett 259:86–90
2. Riddle RD, Johnson RL, Laufer E, Tabin C (1993) Sonic hedgehog mediates the polarizing activity of the ZPA. Cell 75:1401–1416
3. Yokouchi Y, Sasaki H, Kuroiwa A (1991) Homeobox gene expression correlated with the bifurcation process of limb cartilage development. Nature 353:443–445
4. Bruneau BG, Bao ZZ, Tanaka M, Schott JJ, Izumo S, Cepko CL, Seidman JG, Seidman CE (2000) Cardiac expression of the ventricle-specific homeobox gene Irx4 is modulated by Nkx2-5 and dHand. Dev Biol 217:266–277

Check for updates

Chapter 13

Chromosome Painting of Mouse Chromosomes

Lisa L. Hua and Takashi Mikawa

Abstract

Chromosome painting enables the visualization of chromosomes and has been used extensively in cytogenetics. Chromosome paint probes, which consist of a pooled composite of DNA-FISH probes, bind to nonrepetitive sequences for individual chromosomes [1, 2]. Here we describe the process of using chromosome paint to study the organization of chromosomes without fragmenting the nucleus. This method can be used to analyze chromosome position, and identify translocations and ploidy within the nucleus. The preservation of nuclear morphology is crucial in understanding interchromosomal interactions and dynamics in the nucleus during the cell cycle.

Key words Chromosome paint, DNA-FISH, Fluorescence in situ hybridization, Nuclear organization, Nuclear structure, Chromosomes, Cytogenetics

1 Introduction

The field of nuclear organization is rapidly expanding and many unanswered paradigms have come to light. Increasing research supports that chromosome architecture has significant implications during development [3–5]. Common methods of chromosome identification such as conventional banding, or karyotyping, are invasive as it requires rupturing nuclear membranes and can only be applied to mitotic cells. Thus, the consequential analysis of metaphase spreads lacks the spatial information necessary to address native chromosome organization.

Chromosome painting is a powerful technique that can be used to study the topological organization of chromosomes at any cell cycle stage, without nuclear fragmentation. Whole chromosome probes are differentially labeled, specific to nonrepetitive sequences, so each mouse chromosome can be identified [1, 2]. Simultaneous staining of multiple chromosome paint probes, each conjugated with a different fluorophore, allows for visualization for up to three chromosomes along with a nuclear counterstain (Fig. 1). The information can then be used to assess the chromosome identity and relative position in many different cell types. In addition to chromosome

Paul Delgado-Olguin (ed.), *Mouse Embryogenesis: Methods and Protocols*, Methods in Molecular Biology, vol. 1752, https://doi.org/10.1007/978-1-4939-7714-7_13, © Springer Science+Business Media, LLC, part of Springer Nature 2018

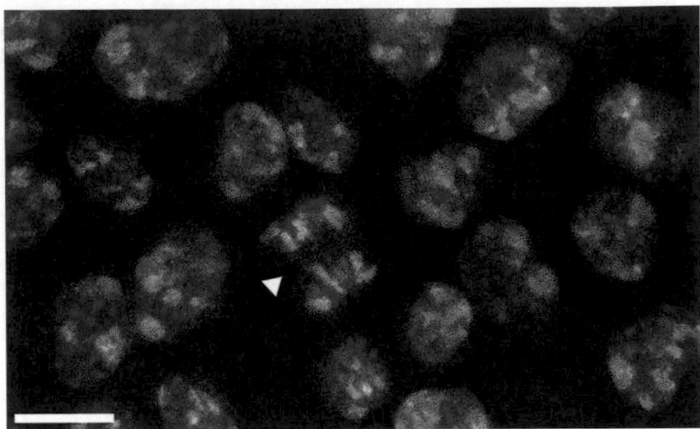

Fig. 1 Chromosome paint labeling in mouse C212 tetraploid myoblasts. Top view of stacked confocal optical sections in spun down nuclei show three sets of duplicated homologous chromosome pairs (chromosome 3 in green, 6 in yellow, and 1 in red) localized in the nucleus (TO-PRO3, a nuclear stain for DNA presented in blue). Chromosome paint probes hybridize to the majority of each chromosome by sequence specificity. During anaphase (arrowhead), the homologs chromosome replicate and each daughter nucleus contain two copies of chromosome 3, 1, and 6. Note: Condensed chromosomes at anaphase stain more intensely then interphase nuclei, which exhibit more diffuse staining throughout the nucleus. Scale bar: 10 μm

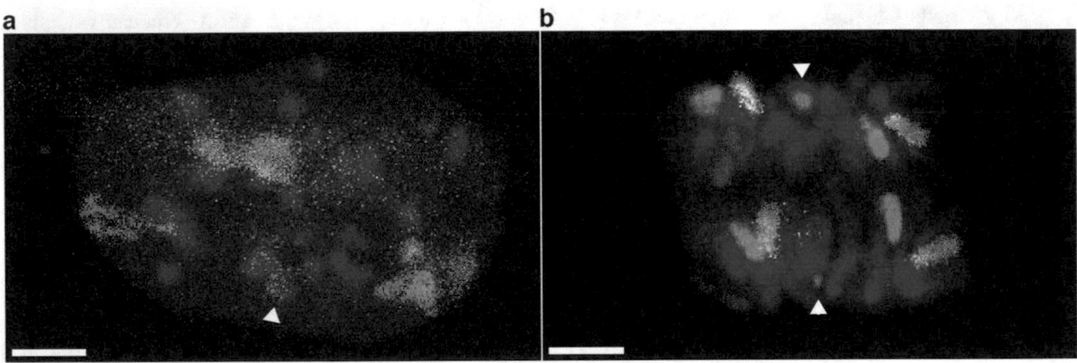

Fig. 2 Supernumerary translocated chromosome in male Ts65Dn [6, 7] mouse can be detected based on size [8, 9]. Top view of stacked confocal optical sections of nuclei (TO-PRO3, a nuclear stain for DNA presented in blue) of primary MEFs of male Ts65Dn mice painted for chromosome X (red), 16 (green), and 17 (yellow) in (**a**) interphase and (**b**) anaphase nuclei. The translocated chromosome 17^{16} (green, arrowhead) can be visualized based on its size. Scale bar: 3 μm

position in the nucleus, chromosome translocations and changes in ploidy can be identified. For example, supernumerary chromosomes can be stained and identified based on their size (Fig. 2).

Currently there are commercially available probes for all murine chromosomes, which provide a technical convenience to many researchers in the field. Chromosome paint probes can be labeled with various fluorochrome–hapten conjugates and imaged on a microscope with proper filters. Here, we demonstrate an approach to stain individual mouse chromosomes, without comprising nuclear morphology, during multiple stages of the cell cycle. A discussion of critical parameters such as chromosome paint probe accessibility, fixation, and pretreatment methods will be necessary as this will impact the protocol's success. It is also reasonable to discuss denaturation, prehybridization/posthybridization steps in detail, as these are essential steps required for successful chromosome painting.

To date, a detailed understanding of how nuclear organization regulates developmental events is limited. It will be critical to compare nuclear organization and dynamics for embryonic cells undergoing their developmental processes of specification, determination, and differentiation.

2 Materials

Prepare all solutions using double deionized water and molecular biology grade reagents. Prepare all reagents the same day of usage and store all reagents at room temperature (unless indicated otherwise). Diligently follow all waste disposal regulations when disposing of waste materials.

2.1 Harvesting Embryonic Fibroblast Cells (MEFs) from Mouse Embryos

1. 1–2 pregnant female mice.
2. 70% ethanol. Use reagent grade ethanol and dd H_2O.
3. 1× PBS: 8 g of NaCl, 0.2 g of KCl, 1.44 g of Na_2HPO_4, 0.24 g of KH_2PO_4. Add to 800 mL of dd H_2O and use stir bar to help with dissolving. Adjust the pH to 7.4 using HCl. Adjust the final volume to 1 L with additional dd H_2O. Sterilize by autoclaving and store at RT.
4. 1× DPBS (magnesium and calcium free PBS).
5. 0.25% trypsin–EDTA solution.
6. 0.05% trypsin–EDTA solution or Accutase.
7. MEFs culture medium: 10% fetal bovine serum, 0.1 mM β-mercaptoethanol, 50 U of penicillin, 50 μg/mL streptomycin in Dulbecco's modified Eagle's medium (DMEM). Use a bottle top filter vacuum filter system (0.22 μm, Corning) to remove impurities from the solution. This solution can be stored at 4 °C for up 2 months.
8. Sterile dissecting tools. Forceps and scissors (Dumont #5).
9. 15 mL and 50 mL sterile centrifuge tubes.
10. Cell strainer (100 μm Nylon).
11. 150 mm or 100 mm cell culture dish.
12. Nitrile gloves.

2.2 Cell Preparation of Adherent Cells

1. 8 mm custom microscope slides (Azerscientific).
2. 150 mm or 100 mm cell culture dish.
3. MEFs culture medium.

2.3 Cell Preparation of Suspension Cells

1. 0.05% trypsin–EDTA solution or Accutase.
2. Cytofunnels, adaptors, and filter paper for Shandon 4 Cytospin.
3. Superfrost slides.
4. MEFs culture medium.

2.4 Fixation and Permeabilization

1. Fixation solution (4% PFA): For a 50 mL solution: 2 g of paraformaldehyde in 50 mL of 1×PBS. Use stir bar and heat (no higher than 65 °C to prevent boiling of PBS) to dissolve paraformaldehyde pellets/powder. Slowly add NaOH until solution is clear. Let solution cool after mixture is clear. Adjust the pH to 7.4 using HCl. Filter with Whatman paper or centrifuge tube filter to remove impurities from the solution. This solution works best if made the day of usage and can be stored at 4 °C for up to 1 week.
2. Permeabilization Solution (1× PBST): 0.5% Tween 20 in 1×PBS. Invert 5–10 times to mix. Make fresh.
3. 70% ethanol. Use reagent grade ethanol and dd H_2O.
4. Coplin jars.
5. Belly Dancer or a general platform rocker.

2.5 Chromosome Paint Protocol

1. Denaturation buffer (2× SSC): 175.3 g NaCl, 88.2 g of sodium citrate in 800 mL of dd H_2O. Use a stir bar to help with dissolving. Adjust the pH to 7.0 using HCl. Adjust the final volume to be 1 L with additional dd H_2O. Sterilize by autoclaving. Dilute down to 2× SSC by adding dd H_2O.
2. 0.1 N HCl.
3. 1× PBS (prechilled at 4 °C).
4. Ethanol. Use reagent grade ethanol and dd H_2O to make appropriate dilutions for ethanol series (70%, 80%, and 100% and prechill at 4 °C).
5. Denaturation solution (preheated): 70% formamide diluted in 2× SSC. Adjust the pH to 7.0 using HCl. Can be made in bulk and stored at 4 °C.
6. Post-hybridization wash buffer (preheated): 55% formamide, 0.1% NP-40 in 2× SSC. Invert 5–10 times to mix.
7. Coplin jars.
8. Mini coverslips (8 mm di, World Precision Instruments)
9. Large coverslips (12CIR.-1, Fisher Scientific).
10. Slide warmer set at 37–42 °C.

11. Rubber cement (Elmer's).
12. Hybridization oven.
13. Forceps (Dumont #5, Fine Science Tools).
14. Belly Dancer (Denville Scientific) or a general platform rocker.

2.6 Chromosome Paint Probes

1. Chromosome paint probes (Applied Spectral Imaging) conjugated to FITC, Rhodamine, or Aqua.
2. 1.5 microcentrifuge tube.
3. Two Heat blocks set at 80 °C and 37 °C.

2.7 DNA Counterstain and Mount

1. DNA stain: 4′,6-diamidino-2-phenylindole (DAPI) or TO-PRO-3 (Thermo Fisher Scientific).
2. Mounting medium: Antifade (Thermo Fisher Scientific) or 9:1 glycerol/1× PBS with pH adjusted to 8.5–9.0.

3 Methods

3.1 Harvesting Primary Mouse Embryonic Fibroblasts

All cell culture solutions and supplies must be sterile and be used under a laminar flow hood. Aseptic technique should be practiced at all times. Cells should be incubated in a humidified 37 °C, 5% CO_2 incubator.

1. Euthanize the pregnant dam at day E13.5 by CO_2 asphyxiation, followed by manual cervical dislocation, according to your institutional guidelines provided by IACUC.
2. Place pregnant dam ventral/belly side up on a few paper towels. Spray the abdominal area and lower half of the pregnant mouse with 70% ethanol.
3. Use nitrile gloves for the remaining steps.
4. Use dissection forceps to pinch the skin of the abdominal wall and lift up. Use the dissection scissors in your other hand to cut a small incision from the abdominal wall in the direction along the anterior–posterior axis.
5. Use both hands to pull/peel the skin from the abdominal area to expose the uterus containing the embryos.
6. Use clean forceps with one hand to lift up the bottom of the uterus, and use clean scissors, with the other hand, to remove the entire uterus containing the embryos.
7. Transfer the uterus containing embryos into a clean 100 mm culture dish containing 1× PBS.
8. Use forceps and scissors to separate individual embryos in the uterus. Transfer the individual embryos into a new 100 mm culture dish containing 1× PBS.

9. Under a dissecting/stereomicroscope, working with one embryo at a time, open the yolk sac and dissect out the embryos. Using forceps, remove and discard the liver, heart, and brain. If embryos have different genotypes, transfer each embryo to individual 15 mL centrifuge tubes. Keep them on ice until all embryos are dissected. Transfer all yolk sacs to 1.5 mL centrifuge tubes for DNA extraction. These will be used for genotyping by PCR [10] or Southern blot [11].
10. Use forceps to fragment embryos into 2–3 pieces each. Transfer embryos to a 50 mL centrifuge tube with up to 30 mL of cold 0.25% trypsin–EDTA (or 3 mL per embryo) on ice.
11. Incubate the 50 mL centrifuge tube containing embryos at 4 °C overnight allowing for trypsin-EDTA to permeate through the embryo with minimal enzymatic activity.
12. The next day, without contacting the embryos at the bottom of the centrifuge tube, aspirate the trypsin–EDTA solution until approximately double the volume of embryonic tissue remains.
13. Incubate the centrifuge tube in a 37 °C water bath for 30–45 min. Vortex the tube every 5–10 min to fragment the tissue and help disassociate cells.
14. In a laminar flow hood, add up to 20–30 mL of MEF culture medium to the 50 mL centrifuge tube containing fragmented embryos, and pipette up and down with a serological pipet repeatedly to aid in disassociating the cells from the tissue.
15. Use a serological pipet to transfer your tissue suspension through a cell strainer fitted over a new 50 mL centrifuge tube. Collect the supernatant as this is your cell suspension. Note: Transfer 1–3 mL at a time as to not overflow the cell strainer. Additionally not fitting the cell strainer tightly on the 50 mL centrifuge tube, rather at a slant, will help speed up the filtration process.
16. Split the cell suspension in the supernatant into 150 mm culture dishes. Divide total supernatant by number of embryos. The ratio is one embryo per 150 mm dish containing 16–18 mL of MEF culture medium.
17. Incubate the MEF cells overnight in a humidified 37 °C, 5% CO_2 incubator.
18. Remove MEF medium and wash cells twice with 1× DPBS to remove nonadherent cells and debris. Note: DPBS, rather than PBS, must be used as Mg^{2+} and Ca^{2+} can inhibit trypsin/Accutase activity.
19. Add enough 0.05% Trypsin-EDTA solution to cover all cells (typically 2–3 mL for a 150 mm cell culture dish) and place in a 37 °C, 5% CO_2 incubator for approximately 5–10 min, or until most cells detach from cell culture plate. Cell morphology will change from spindle-like to a more rounded, shrunken

shape. Alternatively, Accutase can be used instead of trypsin. Cells with Accutase can be incubated at RT for up to an hour.

20. Add 20 mL of MEF media to culture dish to stop neutralize reaction. Pipette the media-containing cells to a new 50 mL centrifuge tube and centrifuge at 300 × *g* for 3–5 min to pellet cells.
21. Aspirate media while not perturbing the cell pellet at bottom. Resuspend cell pellet with 1–5 mL of media and add to cell culture dish containing 30–40 mL of media with submerged slides (flame-sterilized). Additionally, cells can be further expanded in culture or cryopreserved in liquid nitrogen.

3.2 Cell Preparation of Adherent Cells

1. Flame-sterilize (with option of additional UV sterilization) 1–5 slides.
2. Place sterilized slides in cell culture dish and submerge with appropriate medium.
3. Culture adherent cells (*see* above for primary MEFs) on 8 mm slides until they are about 75%–80% confluent.

3.3 Cell Preparation of Suspension Cells

1. Culture cells until they are about 75%–80% confluent.
2. Collect cells in centrifuge tube and spin at appropriate time/speed (optimal centrifuge differs by cell type) to form cell pellet.
3. Aspirate supernatant and resuspend pellet in 500–1000 μL of medium. Keep cells on ice.
4. Add 30–250 μL of cells to each Cytofunnel and use a Cytospin to spin down cells at 800 RPM, or 139 × *g*, for 3 min on Superfrost slides.

3.4 Fixation and Permeabilization

1. Submerge slides in a Coplin jar containing 50 mL of freshly made fixation solution: 4% paraformaldehyde for 10 min at RT. Note: PFA is the recommended fixative as it preserves the nuclear structure the best [12].
2. Remove fixation solution. Wash slides with permeabilization solution (PBST 0.5%Tween 20) in Coplin jar while rocking on a Belly Dancer or a general platform rocker for 10 min at RT.
3. Dehydrate slides with 70% ethanol.
4. Place Coplin jar with slides in ethanol at 4 °C overnight to age slides (Slides can be stored for up to 3 months at −20 °C).

3.5 Chromosome Painting

1. Probe preparation: Heat-denature Whole Chromosome Paint in a 1.5 microcentrifuge tube in a heat block set at 80 °C for 10 min, then cool to 37 °C for ~60 min until use in **step 10**. Note: Multiple chromosome probes can be combined for a single application in the hybridization step.

2. Rehydrate slides in 2× SSC for 10 min on ice, on a Belly Dancer or a general platform rocker.
3. Depurinate slides with 0.1 N HCl for 5 min at RT to fragment DNA. Note: Depurination is an essential step to increase accessibility for chromosome paint probes to their target chromosome.
4. Wash slides with cold 1× PBS for 2–5 min each, with rocking. Repeat wash two additional times.
5. Dehydrate slides in a cold ethanol series: 70%, 80%, and 100% for 2–5 min each on ice, with rocking.
6. Pour out ethanol and air-dry slides in Coplin jar for 1 min at RT.
7. Place slides on slide warmer until slides are dry (~4 min). Note: Glass slides will absorb a large amount of heat of the denaturation solution immediately after submersion and thus decreases the temperature of denaturation solution. To minimize this temperature change, slides must be prewarmed.
8. Emerge slides in prewarmed denaturation solution at 80 °C for 7.5 min in a hybridization oven (temperature and time will vary and cell-type specific).
9. Following heat denaturation, immediately dehydrate slides with a second series of ice cold ethanol washes: 70%, 80%, and 100% for 2–5 min each on ice, while rocking.
10. Pour out ethanol and air dry slides in Coplin jar for 1 min at RT.
11. Place slides on slide warmer until slides are dry (~4 min).
12. Apply 2.5 μL of the Whole Chromosome Paint mixture to the center of each 8 mm well on slide, 2 wells per slide without any air bubbles. Immediately use a pair of forceps to carefully place a mini 8 mm coverslip over each well to seal the sample. Note: If there are air bubbles, or insufficient covering of the cells with chromosome paint, there will be areas devoid of staining when slide is imaged.
13. Overlay a large coverslip over the mini 8 mm coverslips and seal the edges with 700 μL of rubber cement with a trimmed pipette tip. Note: Cut the bottom of a 1000 μL tip to pipette rubber cement as it is a viscous aqueous solution.)
14. Place coverslipped slides at 80 °C for 7.5 min in a hybridization oven (temperature and time will cell-type dependent) to codenature both chromosome probe and sample.
15. Place denatured slides in a humidified chamber and incubate at 37 °C overnight. Note: In situ humidified chambers can be made by using pipette tip boxes or slide boxes that can be sealed/closed containing moistened paper towels in a 50% formamide, diluted in dd H_2O, solution.

16. The following day, carefully remove both large/small coverslip by breaking the sealed edge of rubber cement with forceps. Slowly and carefully peel off, with the forcep tip slightly underneath the large coverslip edge, both coverslips simultaneously. Note: the small coverslip will be adhered to the large coverslip. If the small coverslip remains after the large coverslip is peeled off, carefully submerge the slide in wash solution a few times until it comes off or remove with coverslip with forceps.
17. Immediately following coverslip removal, submerge slides in a Coplin jar containing preheated posthybridization Wash buffer for at 42 °C for 10 min in a hybridization oven, while rocking. Repeat two times.
18. Incubate slides with a nuclear counterstain: DAPI or TO-PRO3 for 5 min at RT.
19. Washed slides in Coplin jar containing 1× PBS for 5 min at RT, while rocking.
20. Use forceps to remove any trace amounts of rubber cement. Dry areas around the wells with Kim Wipes. Mount wells on the slide with Prolong Antifade and large coverslips.
21. Seal coverslips with nail polish.

4 Notes

4.1 Critical Parameters and Troubleshooting

A common problem in chromosome painting is low/or weak signal detected over background staining. This can be caused by the inaccessibility of the probe to its target chromosome by insufficient denaturation. To correct for this, the user must optimize denaturation temperature and time. A general practice would be to start with the recommended times used in this protocol for MEFs then increase either time or temperature to find the optimal conditions for the cells. Certain cells, such as murine cell lines or cardiomyocytes, display robust cell membranes and cytoskeletal elements that require longer denaturation times for probe accessibility to the nucleus. However, longer hybridization times will affect the recovery of the sample as many cells will be lost during this process. Annealing temperatures have been reported at temperature of 72 °C from manufacturer (Applied Spectral Imaging); therefore the user can balance increasing hybridization duration with lowering temperatures for maximum retention of sample.

Another solution to correct for insufficient denaturation is to increase hybridization temperature (up to 85 °C). This will greatly compromise the cellular and nuclear membrane architecture making it easier for the probes to gain access to their targets. However, this can lead to overdenaturation of the nucleus where nuclear morphology is distorted and DNA fragmentation will occur. Consequently loss of cells will also occur.

It is extremely important to prewarm your slides as glass slides absorb heat from the denaturation solution and can lower its temperature, therefore compromising denaturation.

Increasing concentration of HCl can also aid in increasing signal strength as this will depurinate a larger amount of DNA and strip more proteins to allow for easier accessibility for probe to DNA. However this can lead to overfragmentation of chromosomes and provide inconclusive results.

A cause for weak or inconsistent signals could be due to the age of the cell samples on the slide. Older slides can sometimes require longer denaturation times for good signal recovery. However if samples were not kept cold (−20 °C) or in proper conditions, DNA degradation could occur therefore making it inadequate for chromosome painting. Slides made from fresh samples prepared overnight are optimal for chromosome paint protocols.

If there is too much background staining in your nucleus from your chromosome paint, you can alter the stringency of the posthybridization wash solution by decreasing the concentration of SSC and increasing the concentration of formamide and NP-40. This will increase the stringency of your probe to target association and remove other nonspecific DNA binding. Alternatively, too much background staining in the cytoplasm can be removed by pepsin/trypsin/proteinase K digestion as pretreatment for slides.

Acknowledgments

We would like to thank Sara Venters and all the other members of the Mikawa lab for helpful discussions on technical details of the protocol. We would also like to thank Karen Leung (Barbara Panning lab), Eirene Markenscoff-Papadimitriou (Stavros Lomvardas lab), Kiichiro Tomoda (Shinya Yamanaka lab), and Stephanie Parker (Maximiliano D'Angelo lab) for helpful discussions. This work was supported by NIH grant R37HL078921, R01HL112268, R01HL122375, and R01HL132832 to T.M.

References

1. Rabbitts P, Impey H, Heppell-Parton A et al (1995) Chromosome specific paints from a high resolution flow karyotype of the mouse. Nat Genet 9(4):369–375. https://doi.org/10.1038/ng0495-369
2. Ried T, Schrock E, Ning Y et al (1998) Chromosome painting: a useful art. Hum Mol Genet 7:1619–1626
3. Borsos M, Torres-Padilla M (2016) Building up the nucleus: nuclear organization in the establishment of totipotency and pluripotency during mammalian development. Genes Dev 30(6):611–621. https://doi.org/10.1101/gad.273805.115
4. Schneider R, Grosschedl R (2007) Dynamics and interplay of nuclear architecture, genome organization, and gene expression. Genes Dev 21:3027–3043. https://doi.org/10.1101/gad.1604607
5. Mayer R, Brero A, von Hase J et al (2005) Common themes and cell type specific variations of higher order chromatin arrangements

in the mouse. BMC Cell Biol 6:44. https://doi.org/10.1186/1471-2121-6-44

6. Epstein CJ, Cox DR, Epstein LB (1985) Mouse trisomy 16: an animal model of human trisomy 21 (down syndrome). Ann N Y Acad Sci 450:157–168. https://doi.org/10.1111/j.1749-6632.1985.tb21490.x
7. Akeson EC, Lambert JP, Narayanswami S et al (2001) Ts65Dn -- localization of the translocation breakpoint and trisomic gene content in a mouse model for down syndrome. Cytogenet Cell Genet 93(3–4):270–276
8. Reeves RH, Irving NG, Moran TH et al (1995) A mouse model for down syndrome exhibits learning and behaviour deficits. Nat Genet 11(2):177–184. https://doi.org/10.1038/ng1095-177
9. Duchon A, Raveau M, Chevalier C et al (2011) Identification of the translocation breakpoints in the Ts65Dn and Ts1Cje mouse lines: relevance for modeling down syndrome. Mamm Genome 22:674–684. https://doi.org/10.1007/s00335-011-9356-0
10. Green EK (2002) Allele-Specific Oligonucleotide PCR. In: PCR mutation detection protocols, Series: methods in molecular biology, vol 187. Humana Press Inc., Totowa, pp 47–50. https://doi.org/10.1385/1-59259-273-2:047
11. Gebbie L (2013) Genomic southern blot analysis. In: Cereal genomics, Series: methods in molecular biology, vol 1099. Humana Press Inc., Totowa, pp 159–177. https://doi.org/10.1007/978-1-62703-715-0_14
12. Hepperger C, Otten S, von Hase J, Dietzel S (2007) Preservation of large-scale chromatin structure in FISH experiments. Chromosoma 116(2):117–133. https://doi.org/10.1007/s00412-006-0084-2

Check for updates

Chapter 14

Chromatin Immunoprecipitation in Early Mouse Embryos

Estela G. García-González, Bladimir Roque-Ramirez, Carlos Palma-Flores, and J. Manuel Hernández-Hernández

Abstract

Epigenetic regulation is achieved at many levels by different factors such as tissue-specific transcription factors, members of the basal transcriptional apparatus, chromatin-binding proteins, and noncoding RNAs. Importantly, chromatin structure dictates the availability of a specific genomic locus for transcriptional activation as well as the efficiency with which transcription can occur. Chromatin immunoprecipitation (ChIP) is a method that allows elucidating gene regulation at the molecular level by assessing if chromatin modifications or proteins are present at a specific locus. Initially, the majority of ChIP experiments were performed on cultured cell lines and more recently this technique has been adapted to a variety of tissues in different model organisms. Using ChIP on mouse embryos, it is possible to document the presence or absence of specific proteins and chromatin modifications at genomic loci in vivo during mammalian development and to get biological meaning from observations made on tissue culture analyses. We describe here a ChIP protocol on freshly isolated mouse embryonic somites for in vivo analysis of muscle specific transcription factor binding on chromatin. This protocol has been easily adapted to other mouse embryonic tissues and has also been successfully scaled up to perform ChIP-Seq.

Key words Chromatin immunoprecipitation, Embryo, Gene regulation, Epigenetics, Myogenesis, Somites, MyoD, Myogenin

1 Introduction

The study of many fundamental cellular functions such as chromosome segregation, DNA replication, transcription and epigenetic regulation, relies on the analysis of interactions between proteins and DNA in the context of living cells. The chromatin status and the in vivo interaction of transcription factors with DNA as well as the composition of post-translational modifications of histones present at regulatory elements of target genes are part of the dynamic process that determine cellular decisions such as self-cell renewal, proliferation and differentiation during embryonic development [1]. The expression of a gene in a given cell type depends on the interaction of transcription factors with their regulatory sequences and on the acquisition of an appropriate and permissive

Paul Delgado-Olguin (ed.), *Mouse Embryogenesis: Methods and Protocols*, Methods in Molecular Biology, vol. 1752,
https://doi.org/10.1007/978-1-4939-7714-7_14,

chromatin architecture [1–3]. Over the past years, major advances in high-throughput nucleic acids sequencing technologies have increased our ability to identify sequences to which transcription factors and cofactors are bound in vivo and thus, have allowed a better understanding of regulatory networks involved in several developmental processes [4–6]. However, the success of such high-throughput technologies depends on the quality of preparations and starting material.

Chromatin immunoprecipitation (ChIP) has become a technique of choice to dissect the composition of transcriptional regulators in tissue culture cells and in primary tissue [7–9]. Therefore, the applications of ChIP to mouse embryonic tissue provide numerous possibilities to dissect regulatory networks at specific stages and at in vivo contexts that otherwise are difficult to be addressed in tissue culture systems. This, in combination with deep-sequencing methodologies increases the potential and the informative power of ChIP-based methods. However, the application of ChIP to rodent embryonic tissue has proved to be complicated by the limited amount of tissue and the heterogeneity of cell and tissue types in the embryo, especially at early stages of development. Here, we present a method to perform ChIP-qPCR using dissected somites from mouse embryos at embryonic day 9.5 (E9.5). This approach is useful to examine transcription factor binding at skeletal muscle specific gene promoters [10, 11]. Somites, which are the precursor tissues from which the skeletal muscles of the trunk and limbs will form, are present in the mouse at E8.5–9.5 and at this developmental stage skeletal muscle genes such as *MyoD* and Myogenin are actively expressed [12–14]. Sheared chromatin from combined E8.5 embryonic somites can be divided into up to five aliquots, which provides the investigator sufficient material for controls and for investigation of specific protein–chromatin interactions by qPCR. DNA recovery yields after immunoprecipitation are increased by the addition of nucleic acid carriers to minimize material loose. Finally, real-time qPCR assessment of enrichment can be performed on up to five different genomic regions of interest. On follow-up applications, we have used this method to gain input about the biological significance of our ChIP observations made in tissue culture analyses and have successfully generated libraries for next-generation DNA sequencing.

2 Materials

All material must be DNA-free, nuclease-free, and free of any plasmid or PCR products. All the equipment needs to be decontaminated before use, and the use of clean gloves and filter tips and low-retention material tubes is highly recommended for all steps (*See* **Note 1**).

2.1 Reagents

1. 37% formaldehyde.
2. 1.25 M glycine stock solution.
3. 95% (vol/vol) ethanol solution.
4. Phenol–chloroform–isoamyl alcohol.
5. Protease inhibitors cocktail: 1 mM PMSF, 100 μM leupeptin, 0.3 μM aprotinin, 1 μM pepstatin A, 10 μM pepstatin, and 1 mM sodium butyrate.
6. 20 mg/ml Proteinase K stock solution.
7. 3 M sodium acetate stock solution.
8. 1% agarose gel.
9. 1Kb DNA ladder.
10. Glycoblue nucleic acids carrier.
11. 5 M NaCl stock solution.
12. 10% SDS stock solution.
13. Dyneabeads Protein A or protein G (*see* **Note 2**).
14. 1 M Tris–HCl stock solution, pH 8.0.
15. Linear acrylamide nucleic acids carrier.
16. Phosphate buffer saline (PBS); Sterile solution.
17. Antibodies of choice (Use ChIP-grade or IP-grade antibodies when available). Rabbit anti-MyoD (Santa Cruz, sc-304), rabbit anti-Myogenin (Santa Cruz, sc-576) normal rabbit IgG (Millipore, 12–370).
18. qPCR-SYBR green master mix.
19. Mouse Myogenin promoter forward oligo: 5'-CAGGC AGGAGCACGGCAGAC-3′; Myogenin promoter reverse oligo: 5'-ACACAGCCAGGCGTTCACTCC-3′. Mouse IgH enhancer promoter forward: 5'-GCCGATCAGAACCAGAA CACC-3′, IgH enhancer promoter reverse: 5'-TGGTGGGG CTGGACAGAGTGTTTC-3′.

2.2 Buffers and Solutions

1. Dissection medium: Dulbecco's Modified Eagle's Media-DMEM, 10% fetal bovine serum, 20 mM HEPES pH 7.4, 60 μg/ml penicillin, 2 mM streptomycin.
2. Collagenase solution: 100 U/ml collagenase type II in 1× Dubecco's Phosphate Buffered Saline (DPBS).
3. Lysis buffer: 50 mM Tris–HCl pH 8.1, 10 mM EDTA, 1% SDS. Supplemented with protease inhibitor cocktail before using.
4. Immunoprecipitation buffer: 16.7 mM Tris–HCl pH 8.1, 1.2 mM EDTA, 1.1% Triton X-100, 0.01% SDS, 167 mM NaCl. Supplemented with protease inhibitor cocktail before using.

5. Washing buffer A: 20 mM Tris–HCl pH 8.1, 2 mM EDTA, 1% Triton X-100, 0.1% SDS, 150 mM NaCl. Supplemented with protease inhibitor cocktail before using.
6. Washing buffer B: 20 mM Tris–HCl pH 8.1, 1% Triton X-100, 0.1% SDS, 500 mM NaCl. Supplemented with protease inhibitor cocktail before using.
7. Washing buffer C: 10 mM Tris–HCl pH 8.1, 1 mM EDTA, 1% NP-40, 1% sodium deoxycholate, 0.25 M LiCl. Supplemented with protease inhibitor cocktail before using.
8. TE buffer: 10 mM Tris–HCl, pH 8.0, 2 mM EDTA. Supplemented with protease inhibitor cocktail before using.
9. Elution buffer: 0.1 mM $NaHCO_3$, 1% SDS.

2.3 Laboratory Equipment

1. Ultrasonicator Covaris-S220, with a 6 mm microtube adaptor chamber.
2. Covaris microtube, Pre-slit Snap-cap (6 × 16 mm).
3. Micropipets and filtered pipette tips (10 μl, 200 μl, and 1000 μl).
4. Magnetic holder for 1000 μl tubes.
5. 0.6 and 1.5 ml centrifuges tubes.
6. Microcentrifuges.
7. Water bath.
8. Fixed rotation mixer.
9. Heat blocks.
10. Vortex.
11. Real-time-qPCR machine.
12. Photodocumentation system for agarose gels.

3 Methods

3.1 Isolation of Embryos

1. Before conducting any experimental procedure on mice embryos, make sure to follow all the animal care, biosafety, and bioethical institutional guidelines.
2. By noon of the day that the mating plug is observed in the female mouse is considered as embryonic day 0.5 (E0.5) of development. This is confirmed by the presence of a mating plug in the female mouse the morning after mating. At E9.5 sacrifice the mouse following an institution-approved protocol.
3. Spray the abdominal area of the euthanized animal with 70% ethanol and open the abdominal cavity. The implantation sites indicating the presence of developing embryos are seen along each uterine horn. Remove both uterine horns, cut each implantation

site individually to separate and place them in a 100 mm petri dish containing dissection medium at room temperature.

4. Isolate individual embryos under a dissection microscope. The visceral yolk sac is located between the amnion and the parietal yolk sac and is readily distinguished in E9.5 embryos by the presence of prominent blood vessels that nurture the embryo. Remove each embryo from its surrounding membranes and transfer individually into a new plate containing dissection media to remove excess blood. Change the media three times to remove the blood and debris.
5. We routinely analyze transcription factors binding at myogenic loci in developing somites. For this, we dissect the rostromedial somites from each embryo and collect them in a 1.5 ml tube containing 200 μl of dissection media (Fig. 1a). These structures are detectable under the stereoscopic microscope in embryos from this stage. Combine the somites from three embryos in each tube and proceed with the protocol.
6. As a negative control, we use yolk sac tissue. Dissect yolk sac and collect it in a 1.5 ml tube containing 200 μl of dissection media and process alongside the somite samples.

3.2 Embryo Tissue Homogenization

1. Add 20 units of Collagenase type II in a volume of 200 μl of dissection media to each 1.5 ml tube containing embryonic somites and yolk sac (*see* **Note 3**).
2. Incubate at 37 °C with agitation for 20 min.
3. Resuspend by pipetting up and down to disrupt clumps and to get a cell suspension.
4. Apply the cell suspension on the top of sterile cell strainer (40 μm mesh size) previously rinsed with fetal bovine serum and placed it over a 1.5 ml micro centrifuge tube (*see* **Note 4**).
5. Immediately apply 600 μl of fetal bovine serum at room temperature on top of the cell strainer to complete the separation. Discard the strainer.
6. Centrifuge samples at 4 °C at 1,000 × *g* for 5 min.
7. Discard the supernatant.
8. Resuspend the pellet in 1 ml 1× DPBS/10% fetal bovine serum at room temperature (*see* **Note 5**).
9. Centrifuge the samples at 1,000 × *g* for 1 min at room temperature.
10. Discard the supernatant.
11. Resuspend the pellet in 200 μl dissection media at room temperature (*see* **Note 6**).
12. Count the embryonic cells. There should be an average 3–5 × 10^6 cells in three E8.5 embryos (*see* **Note 7**).

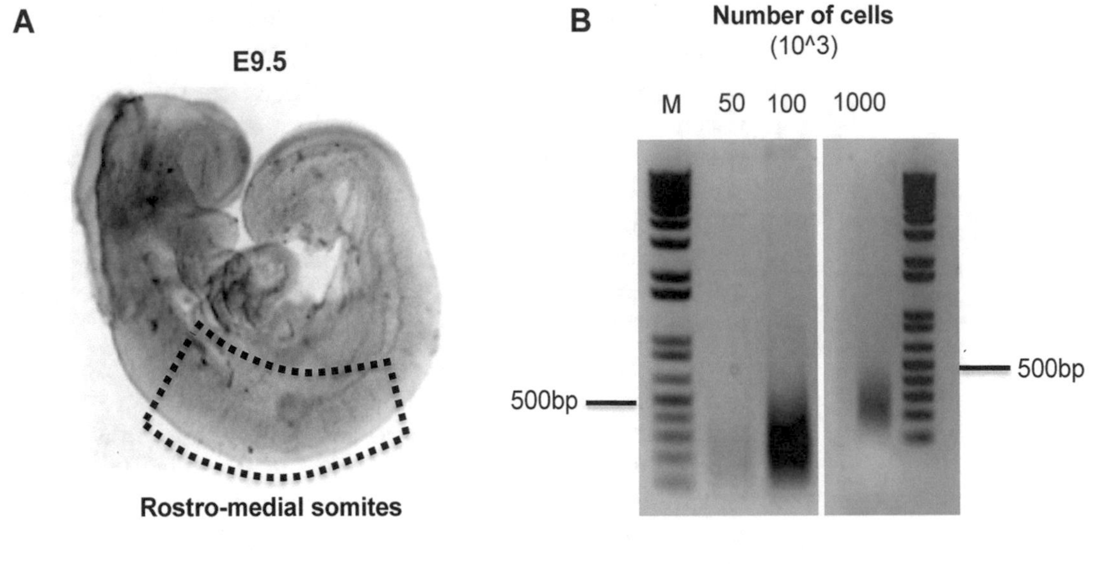

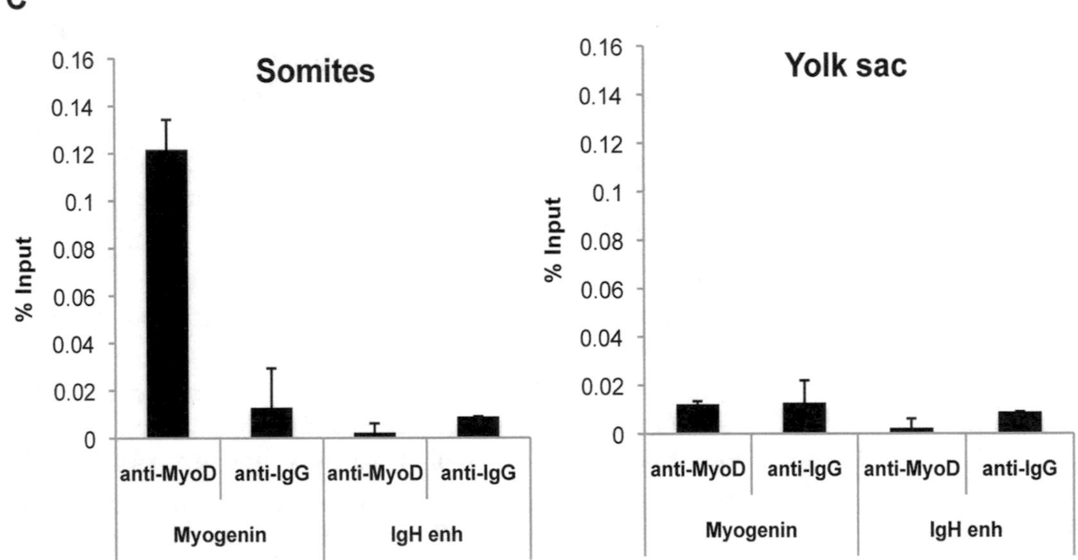

Fig. 1 Representative Results of ChIP on early mouse embryos. (**a**) Dissected E9.5 embryo. The dotted area represents the rostromedial somites that are dissected using a scalpel and collected for further homogenization. (**b**) Representative results of chromatin sonication using different number of cells. A bulk between 200 and 500 bp is expected after reversing cross-link and DNA purification. Purified DNA aliquots are separated in 1% agarose gel. (**c**) Representative results of two independent ChIP-qPCR experiments showing enrichment of MyoD at the Myogenin promoter specifically in dissected somites but not in yolk sac used as negative control. *See* differences in the enrichments between Myogenin promoter and IgH enhancer regions

3.3 Chromatin Cross-Linking

1. Add 5.6 μl of 37% formaldehyde to the 200-μl samples to get a final concentration of 1% formaldehyde.
2. Incubate samples for 10 min at room temperature with gentle but constant agitation.

3. Stop cross-linking by adding glycine to a final concentration of 0.125 M (*see* **Note 8**).
4. Centrifuge the samples at 4 °C at 1,000 × *g* for 3 min.
5. Discard the supernatant.
6. Resuspend the cells with cold 1× DPBS/10% fetal bovine serum.
7. Centrifuge samples at 4 °C at 1,000 × *g* for 3 min.
8. Discard the supernatant. At this step, pellets can be frozen at −80 °C for several weeks.

3.4 Preparation of Antibody-Bead Complexes

1. Prepare slurry of Dynabeads protein-A or protein-G according with the isotype of your antibody (*see* **Note 2**). For each antibody, add 20 μl of Dynabeads stock solution into a 1.5 ml tube, place the tube in the magnetic holder and allow beads to be captured. Remove the supernatant, remove the tube from the magnet and add 500 μl of immunoprecipitation buffer.
2. Vortex the solution for 10 s, capture the beads with the magnetic holder, discard the buffer and add 500 μl of immunoprecipitation buffer. Wash the beads twice.
3. Add 200 μl of immunoprecipitation buffer to the washed beads and add 3 μg of the antibody of interest (*see* **Note 9**).

Do not forget to include a negative control with IgG of the same isotype of the antibody or a no-antibody control. Place the tubes at 50 rpm on a rotator overnight at 4 °C.

3.5 Sonication of Chromatin

1. Resuspend cell pellet from Subheading 3.3, **step 8** in 130 μl of iced-cold lysis buffer and transfer the content to a Covaris sonicator tube. Make sure not to form bubbles and put the tube into the microtube adaptor.
2. Fill the Covaris chamber with bidistilled water and set the following parameters: peak incident power (watts): 105; Duty Factor (percent): 5.0; Cycles/Burts (counts): 200; Duration (seconds): 600.
3. Start the sonication cycle. Keep the rest of the samples on ice.
4. When standardizing chromatin sonication for different samples or cell numbers, add an extra microtube to test sonication conditions and proceed with nucleic acids purification described in Subheading 3.7 to assess DNA shearing on a 1% agarose gel (Fig. 1b); (*see* **Note 10**).
5. Once sonication conditions are set up, proceed with immunoprecipitation.

3.6 Immunoprecipitation and Washes

1. After sonication, transfer 130 μl of soluble chromatin from Subheading 3.5, **step 3** into a 1.5 ml tube and centrifuge for 5 min at 13,000 × *g* to eliminate cellular debris.
2. Add the soluble chromatin to the coupled antibody-beads from Subheading 3.4 in a final volume of 1 ml of immunoprecipitation buffer and mix gently.

   Incubate at 4 °C overnight with rotation.
3. Collect the immunocomplexes with the magnetic holder and keep the supernatant at 4 °C to use as input material.
4. Resuspend the immunocomplexes with 1 ml of ice-cold washing buffer A and incubate for 5 min at 4 °C with constant rotation.
5. Pellet the immunocomplexes with the magnetic holder and eliminate the supernatant. Add 1 ml of ice-cold washing buffer B and incubate for 5 min at 4 °C with constant rotation.
6. Pellet the immunocomplexes with the magnetic holder and eliminate the supernatant. Add 1 ml of ice-cold washing buffer C and incubate for 5 min at 4 °C with constant rotation.
7. Pellet the immunocomplexes with the magnetic holder and eliminate the supernatant. Add 1 ml of room temperature TE buffer and incubate for 5 min with constant rotation. Repeat the TE washing one more time.
8. Pellet the immunocomplexes with the magnetic holder and resuspend with 300 μl of elution buffer for 1 h with constant rotation at room temperature.
9. Spin down the beads at 13,000 × *g* for 5 min and transfer the supernatant to a new 1.5 ml tube. Add 20 μl of 5 M NaCl and incubate at 65 °C from 4 h to overnight.
10. Add 20 μl of 5 M NaCl to the unbound input material from Subheading 3.6, **step 3** and incubate at 65 °C from 4 h to overnight.

3.7 DNA Recovery and Extraction

1. Add the equal volume of phenol–chloroform–isoamyl alcohol to the de-cross-linked chromatin from Subheading 3.6, **steps 9** and **10**. Mix the content for 15 s and centrifuge at 13,000 × *g* for 10 min. Collect the aqueous phase and transfer it to a new 1.5 ml tube (*see* **Note 11**).
2. Add 0.1 equal volume of 3 M sodium acetate, 2 μg of glycogen, 2 μg of linear acrylamide, and 2.5 volumes of 100% ethanol. Mix and incubate from 4 h to overnight at −20 °C (*see* **Note 12**).
3. Centrifuge for 30 min at 13,000 × *g* and wash the pellet twice with 1 ml of 70% ethanol.
4. Resuspend in 50 μl of bidistilled water and store at −20 °C until use.

3.8 Real-Time qPCR and Analysis of Data

1. For each new pair of primers, make sure to optimize amplification conditions so that primer efficiencies are comparable between loci of interest. Prepare a master mix and aliquot for individual 25 μl qPCR reactions as follows: MilliQ water 6.5 μl; SYBR Green Master Mix 12.5 μl; forward primer (10 μM stock) 0.5 μl; reverse primer (10 μM stock) 0.5 μl; DNA template 5 μl for all ChIPs and input samples using each primer pair.
2. Set up a 40-cycle real-time PCR program with a final melting curve step.
3. Acquire data and export into Excel spreadsheets.
4. Determine the amount of precipitated DNA relative to input material following the next formula: $(2^{\wedge(\text{Ct IgG-Ctinput})\_(\text{Ct Ab-Ct input})]})$ [15]; (Fig. 1c).

We analyze at least two biological replicates for ChIPs with triplicate qPCR reactions (*see* **Note 13**).

4 Notes

1. Make sure to use low-retention material tubes and tips. This will reduce the loss of material, especially after collagenase digestion.
2. The type of beads depends of the species and the subclass of the antibody of interest. Use protein A beads with rabbit IgG and protein G beads with mouse IgG.
3. Collagenase must be prepared and aliquot at −20 °C. Significant reduction in its efficiency has been observed during repeated cycles of freezing and thawing.
4. Using fetal bovine serum instead of PBS to rinse the cell strainer will avoid that cells get attached to the mesh, do not rinse with PBS.
5. Addition of fetal bovine serum to PBS allows better recovery of cell pellet after tissue dissection and after cross-linking.
6. It is not recommended to freeze embryos or dissected material before performing formaldehyde cross-linking.
7. The average number of cells may be different between experiments, make sure to split the same number of cells for all the antibodies. We have successfully performed ChIP-qPCR using as low as 10,000 cells.
8. Prepare glycine solution, lysis buffer, and washing buffers fresh for every new ChIP experiment.
9. It is recommended to test the antibody before start with a ChIP experiment. This can be done by performing a western blot to confirm the expected molecular weight of the protein

to be analyzed and a coimmunoprecipitation experiment to confirm the ability of the antibody to recognize the epitope of the target protein in solution. It is also worth testing several concentrations of antibody to obtain the best signal-to-noise ratio.

10. When standardizing chromatin sonication for different samples or cell numbers, add an extra microtube to test sonication conditions and proceed with nucleic acids purification described in Subheading 3.7 to assess DNA shearing on a 1% agarose gel alongside 1 kb DNA ladder. This is important because oversonication results in very small fragments that cannot be PCR-amplified efficiently, whereas too little sonication leads to large fragments that cause a high background. Ideally, the DNA fragments in the sonicated material should range from 150 to 1000 bp in size, with the majority of fragments around 500 bp.
11. Safe-lock micro centrifuge tubes are recommended for the long incubation at 65 °C and the subsequent phenol extraction.
12. The use of glycoblue and linear acrylamide increases the DNA recovery ratio; however it is important to wash extensively the DNA pellet after centrifugation with 70% ethanol as residues can inhibit PCR amplification efficiencies.
13. If high background is observed, increase the number of washes with washing buffer A and B.

References

1. Koster MJE, Snel B, Timmers HTM (2015) Genesis of chromatin and transcription dynamics in the origin of species. Cell 161(4): 724–736
2. Bartholomew B (2014) Regulating the chromatin landscape: structural and mechanistic perspectives. Annu Rev Biochem 83:671–696
3. Benayoun BA, Pollina EA, Brunet A (2015) Epigenetic regulation of ageing: linking environmental inputs to genomic stability. Nat Rev Mol Cell Biol 16(10):593–610
4. O'Neill LP, VerMilyea MD, Turner BM (2006) Epigenetic characterization of the early embryo with a chromatin immunoprecipitation protocol applicable to small cell populations. Nat Genet Jul 38(7):835–841
5. Gilfillan GD, Hughes T, Sheng Y, Hjorthaug HS, Straub T, Gervin K et al (2012) Limitations and possibilities of low cell number chip-seq. BMC Genomics 13:645
6. Liu X, Wang C, Liu W, Li J, Li C, Kou X et al (2016) Distinct features of h3k4me3 and h3k27me3 chromatin domains in pre-implantation embryos. Nature 537(7621): 558–562
7. Tortelote GG, Hernández-Hernández JM, Quaresma AJC, Nickerson JA, Imbalzano AN, Rivera-Pérez JA (2013) Wnt3 function in the epiblast is required for the maintenance but not the initiation of gastrulation in mice. Dev Biol 374(1):164–173
8. Dahl JA, Collas P (2009) MicroChIP: chromatin immunoprecipitation for small cell numbers. Methods Mol Biol 567:59–74
9. Dahl JA, Collas P (2008) MicroChIP-a rapid micro chromatin immunoprecipitation assay for small cell samples and biopsies. Nucleic Acids Res 36(3):e15
10. Hernández-Hernández JM, Mallappa C, Nasipak BT, Oesterreich S, Imbalzano AN (2013) The scaffold attachment factor b1 (safb1) regulates myogenic differentiation by facilitating the transition of myogenic gene chromatin from a repressed to an activated state. Nucleic Acids Res Jun 41(11): 5704–5716

11. Buckingham M, Rigby PWJ (2014) Gene regulatory networks and transcriptional mechanisms that control myogenesis. Dev Cell 28(3):225–238
12. Cho OH, Mallappa C, Hernández-Hernández JM, Rivera-Pérez JA, Imbalzano AN (2015) Contrasting roles for myod in organizing myogenic promoter structures during embryonic skeletal muscle development. Dev Dyn Jan 244(1):43–55
13. Musumeci G, Castrogiovanni P, Coleman R, Szychlinska MA, Salvatorelli L, Parenti R et al (2015) Somitogenesis: From somite to skeletal muscle. Acta Histochem 117(4–5):313–328
14. Pourquié O (2001) Vertebrate somitogenesis. Annu Rev Cell Dev Biol 17:311–350
15. Livak KJ, Schmittgen TD (2001) Analysis of relative gene expression data using real-time quantitative PCR and the 2(−delta delta C(T)) method. Methods 25(4):402–408

Chapter 15

Shaping Up the Embryo: The Role of Genome 3D Organization

Karina Jácome-López and Mayra Furlan-Magaril

Abstract

The spatial organization of the chromatinized genome inside the cell nucleus impacts genomic function. In transcription, the hierarchical genome structure creates spatial regulatory landscapes, in which modulating elements like enhancers can contact their target genes and activate their expression, as a result of restricting their exploration to a specific topological neighbourhood. Here we describe exciting recent findings obtained through "C" technologies in pluripotent cells and early embryogenesis and emphasize some of the key unanswered questions arising from them.

Key words Genome 3D organization, Topologically associated domains, Boundary, Enhancer, Chromatin, Chromosome conformation capture, Embryonic stem cells

1 Introduction

The arrangement of the genome inside the nuclear space has been subject of study for several decades. Microscopy analysis of cells showed that the genome is structured in electrodense regions corresponding to heterochromatin often located at the nuclear periphery and surrounding the nucleoli and light euchromatic regions dispersed in the nucleoplasm [1, 2]. Soon after, FISH technologies showed that chromosomes occupy discrete territories in the interphase nuclei and started exploring how specific genomic regions locate relative to each other and to nuclear bodies [3, 4]. A lot of questions arose from these pioneering observations regarding the influence of genome organization in several genomic processes. In recent years, the development of "C" methods derived from Chromosome Conformation Capture, together with parallel next generation sequencing have provided a powerful platform to visualize how genomes distribute inside the nucleus at high resolution, producing exciting insights and new questions on nucleome

Paul Delgado-Olguin (ed.), *Mouse Embryogenesis: Methods and Protocols*, Methods in Molecular Biology, vol. 1752, https://doi.org/10.1007/978-1-4939-7714-7_15,

function. In this review we summarize the "C" methods available and discuss some of the most recent findings on the role of the 3D genome in pluripotent cells, early embryogenesis and disease.

2 Genome Architecture Through "C" Technologies

2.1 Measuring Chromatin Contacts

3C (from chromosome conformation capture) was the first method described that determined contacts between pairs of genomic sequences by measuring their ligation frequency in a cross-linked environment either by PCR or sequencing [5]. All the other subsequent technologies involve methodological adaptations of 3C and as mentioned, they all infer the proximity between chromatin segments by quantifying the ligation frequency between them (Fig. 1). These methods start with formaldehyde cross-linking of the cell followed by nuclei isolation, chromatin restriction or fragmentation and ligation, except from the native conformation capture protocol in which there is no chemical fixation [6]. 4C (one-to-all) is useful to capture the genome-wide interaction profile from a specific region (commonly called bait, anchor or viewpoint) or multiple regions (up to tens) at once [7–9]. 5C (many to many) resolves contacts between hundreds of fragments within a specific region of interest [10, 11]. Hi-C (all vs all) technology allows identification of genome-wide contacts between all genomic chromatin fragments at once [12]. In Hi-C, a fill-in reaction with a biotinylated nucleotide is performed before ligation. Afterward the biotinylated ligation junctions are pulled down with streptavidin beads and prepared for sequencing (Fig. 1). Hi-C provides the full contact set from the cell population, however, depending on the genome size and the enzyme used to restrict the genome, library complexity can hinder the resolution obtained and very deep sequencing is needed. An alternative to overcome this is found in Capture Hi-C in which the Hi-C library is subjected to hybridization with a collection of biotinylated probes to enrich for the interactomes of multiple baits of interest simultaneously (from one to thousands) [13–15]. Hi-C has also been adapted to work on single cells providing the basis to model chromosome structure, and to elucidate which of the topological features arising from "C" maps arise from the cell population and which ones are present in every single cell [16–19].

The interactions obtained from the techniques above are not associated to a particular protein but represent the ensemble of contacts mediated by the proteins present at the time of cross-linking. ChIA-PET and more recently Hi-ChIP represent an alternative to assess the interactome mediated by a specific protein factor of interest [20, 21]. Both employ a combination of 3C and chromatin immunoprecipitation (ChIP) however in Hi-ChIP the IP is performed after Hi-C minimizing the false positives [21]. So

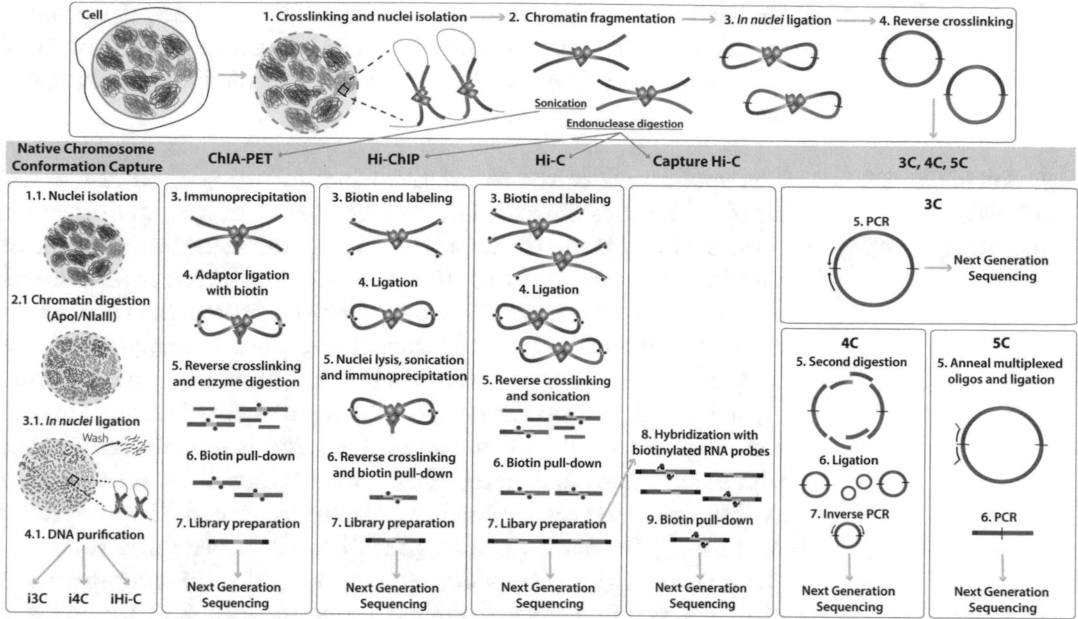

Fig. 1 "C" technologies. The horizontal panel above illustrates the steps for 3C many of which are shared among all techniques. Step 1: Formaldehyde cross-linking and nuclei isolation. Step 2: DNA digestion with a restriction endonuclease; some of the enzymes commonly employed include HindIII, BglII, MboI, and DpnII. Step 3: Ligation of the cross-linked chromatin *in nuclei* to generate hybrid DNA molecules between fragments located in close spatial proximity. Step 4: Cross-link reversal and DNA purification. Step 5: The detection of the DNA contacts can be assessed by PCR using specific primers or Next Generation Sequencing (NGS). In 4C, after cross-link reversal a second endonuclease digestion is made (step 5), followed by religation (step 6); products are then amplified by inverse PCR with bait specific primers and prepared for NGS (step 7). For 5C, multiple specific oligos containing an adapter sequence (red and yellow) for further PCR and library preparation are ligated (step 5) and the resulting molecules analyzed by NGS. Hi-C employs a restriction enzyme that leaves a 5′ overhang, which is filled including a biotin-labeled nucleotide (step 3). After blunt-end ligation (step 4), the Hi-C library is sheared and subjected to pull-down of the biotin-containing fragments (steps 5 and 6) ensuring enrichment of ligation junctions that are subsequently sequenced (step 7). Capture Hi-C begins with a Hi-C library, which is hybridized with biotinylated RNA probes against regions of interest (step 8) which are recovered by a second pull down with streptavidin beads and sequenced. For Hi-ChIP the protocol is the same until HiC step 4, followed by nuclear lysis and sonication. Then, an antibody against a protein of interest is used to immunoprecipitate the occupied specific chromatin fragments (step 5) and the protocol continues as for Hi-C. In ChIA-PET, chromatin fragmentation is achieved by sonication, followed by immunoprecipitation. The free DNA ends in isolated complexes are then complemented with biotin adaptors and ligated (step 4) followed by pull down and library preparation. All methods described above start with formaldehyde cross-linking, in contrast the Native Chromosome Conformation Capture is made without fixation. First the nuclei are isolated (step 1.1) followed by a permeabilization to digest chromatin with ApoI or NlaIII restriction enzymes (step 2.1) and ligation in nuclei (step 3.1). Next, DNA is purified (step 4.1) and prepared to be analyzed by PCR or NGS. This method can be adapted to obtain i3C, i4C, or iHi-C libraries

far, only chromatin contacts mediated by proteins have been revealed. In the years to come it will be crucial to distinguish how other molecules (e.g. RNA) are involved in mediating chromatin interactions.

2.2 Out of the "C": Territories, Compartments, TADs and Loops

The application of the 3C-derived methods described above has unfolded a series of exciting observations about how genomes are organized inside the nucleus. Hi-C has confirmed the existence of chromosome territories so that the contacts between sequences from the same chromosome are much more prominent than across chromosomes [12]. PCA (principal component analysis) of the contact maps generated through Hi-C has shown that the genome is partitioned into two large spatial compartments (expanding several megabases, Mb) of preferred interactions, mostly correlating with euchromatic and heterochromatic regions [12]. Hi-C as well as 5C, also revealed smaller structures called Topologically Associating Domains (TADs) [22, 23]. TADs are self-associating chromatin regions between 0.2 and 1 Mb on average with high contact frequency delimited by boundaries, so that contacts inter-TADs are relatively infrequent [22, 24]. Self-associating domains have been reported in mammals [22], Drosophila [24], *Schizosaccharomyces pombe* [25], in hermaphrodites X-chromosomes of *Caenorhabditis elegans* [26], and in the bacteria *Caulobacter crescentus* [27].

In mammalian genomes the first study reported 2200 TADs with 50–76% conservation in diverse cell types and between mouse and human [22]. More recently, TAD identification at higher resolution showed 9274 TADs in GM12878 human cell line, of which 54% are conserved in other human cell lines [28]. Within TADs, regulatory elements contact their target genes and structural proteins organize their action through insulation and local loops formation. This hierarchical genomic architecture, together with other epigenomic processes, ultimately regulates transcriptional output and its disruption can lead to disease.

What limits the transition between compartments and TADs is not fully understood, however in mammals it is known that TAD boundaries are DNaseI hypersensitive sites rich in housekeeping genes, active histone marks (H3K4me3 and H3K36me3), SINEs sequences and architectural proteins like CTCF and cohesin [22]. A recent report showed that CTCF is essential to preserve TAD integrity in a reversible manner and does not contribute to compartment formation [29]. However, if boundaries delimiting TADs belonging to different compartments are occupied by a particular combination of factors that differ from boundaries separating TADs within the same compartment, remains to be determined. Genetic modifications of certain TAD boundaries lead to alterations in gene expression and in some cases severe phenotypes (*see* below). Recently a distinction between domains and loops has been proposed in Drosophila [30]. However, the precise difference between the two structures is not clear as a TAD can also be thought as a loop with a wider stem. It will be interesting to determine what distinguishes the base of a loop from a TAD boundary.

3 Genome 3D Organization in ESCs: Who Holds the Pluripotent Interactome?

Many studies on genome 3D organization both trough “C” techniques and microscopy have been done using embryonic stem cells (ESCs) aiming to understand the 3D pluripotent genome architecture. Chromatin in ESCs is generally less condensed than in lineage-committed cells and heterochromatin is gained upon cellular differentiation [31–34]. ESCs have structural features found in committed cells including compartments, TADs, and loops. Even if chromatin is generally more relaxed in ESCs and heterochromatin is gained upon differentiation, genome-wide analysis of all promoter interactions in ESCs has revealed that the strongest interactions are found between repressed gene promoters, involved in lineage commitment and early development. In contrast, in a committed cell the most prominent contacts shaping genome architecture, occur among promoters of highly transcribed genes [14]. Defining the topological features providing plasticity to the first cells in the embryo and the ESCs epigenome is currently under extensive research.

3.1 Polycomb Complexes

Polycomb repressive complexes are key transcriptional regulators of developmental genes and they act through the modulation of genome architecture. In Drosophila, Polycomb proteins (PcG) have been described as essential in organizing the genome by recruiting chromatin fibers harbouring target genes into PcG aggregates, silencing their expression [35, 36]. The protein Polycomb (Pc), part of the PRC1 complex, anchors most genomic loops in Drosophila [30], a role that has been mostly assigned to CTCF and Cohesin in mammals. However, as in Drosophila, PcG proteins are fundamental to shape the pluripotent 3D genome.

4C experiments have shown that removal of Eed subunit of PRC2 complex leads to disruption of 3D contacts between regions occupied by H3K27me3 [37]. Promoter Capture-HiC experiments exploring the interactions from all promoters in the genome have shown that the strongest interaction networks in ESCs are formed among early developmental genes [14] and members of the PRC1 complex (Ring1a/b) mediate these contacts [14]. The most prominent interactions are established between the Hox gene clusters, and removal of PRC1 repression leads to promoter network disaggregation. Early lineage specification genes also contact poised enhancers in ESCs and these interactions are not altered upon loss of PRC1 [14]. Recent work has distinguished that is the canonical PRC1 complex the one mediating 3D contacts; 5C performed from PcG-bound regions showed the formation of tightly interacting discrete domains that differ from TADs in ESCs and change upon differentiation to neural progenitor cells (NPCs). These repressive domains are disrupted upon removal of the Polyhomeotic subunit of the canonical PRC1 and interestingly, CTCF is not found at the boundaries of PRC1 domains [38]. It will be interesting to investigate if PcG protein complexes are key players shaping nuclear architecture also in adult somatic cell types and during cellular transformation as reviewed in [39, 40].

3.2 Pluripotency Factors

A group of transcription factors including Sox2, Nanog, Oct4, Klf4, Essrb, and c-Myc are fundamental to maintain the pluripotent state of ESCs [41]. Besides occupying gene promoters, Oct4, Nanog, and Sox2 have been shown to co-occupy, together with Mediator, Klf4, and Essrb, large enhancers (super-enhancers) that control genes involved in maintaining ESC identity and pluripotency [42]. Promoter Capture-HiC analysis has shown that 67% (142 out of 210) of ESCs super-enhancers contact their predicted target gene by linear proximity and they contact an additional set of 361 highly expressed genes also linked to ESC cell identity [14]. However, super-enhancers do not engage in more interactions than a regular enhancer [14]. Super-enhancer requirement for gene expression has been functionally assessed at the *Sox2* locus in which gene editing of its super-enhancer abrogates *Sox2* gene expression by 90% [43].

4C experiments performed from selected promoters as viewpoints together with analysis of HiC data has shown that regions in the genome occupied by clusters of pluripotency factors tend to gather in the nuclear space. Removal of Nanog and Oct4 decreases such contact frequencies; however it does not seem to alter the overall structure of the genome [37, 44]. In particular, confirmation of the role of Klf4 in mediating long-range interactions was described through 4C from the *Oct4* promoter. The promoter of the *Oct4* gene contacts an enhancer through Klf4-mediated cohesin recruitment. Importantly the interaction is formed prior to transcriptional activation during reprograming [45].

So far, many promoter–enhancer and promoter–promoter interactions have been described. Further studies will be needed to address if interactions among enhancers have an important contribution to build their topological landscape.

Both PcG proteins and pluripotency factors are essential to maintain the pluripotent state. It will be important to contrast the effects of alterations in PcG and pluripotency factors regarding nuclear architecture to discriminate if there is a driver and a passenger shaping the pluripotent interactome or if both equally impact the ESC 3D landscape.

3.3 Architectural Proteins

CTCF and cohesin proteins (Rad21, SMC1, SMC3, and STAG) have been recognized as key players mediating chromatin loop formation and boundaries between TADs [46–48]. Rad21 ChIP-seq analysis in ESCs has revealed a subset of cohesin, CTCF independent binding sites, co-occupied by pluripotency transcription factors and removal of Rad21 causes similar defects as Nanog deficient cells indicating cohesin has an important role establishing/maintaining pluripotency [49]. Smc1 ChIA-PET showed that enhancers and genes maintaining pluripotent identity of ESCs are insulated within loops anchored by both CTCF and cohesin [50].

Recently, contradictory results regarding the relevance of CTCF in organizing the ESCs 3D genome using the same technology (Auxin-Inducible Degron (AID) system) to degrade a tagged CTCF, have been reported. Nora et al. observe a severe reversible effect at the level of loops and TAD formation upon CTCF degradation with concomitant defects in gene expression [29]. However, Kubo et al. do not observe a dramatic change at the TAD level and no measurable effects on histone marks or gene expression [51]. The discrepancy between these results might be due to ESC clone usage or the analytical tools implemented. Further experiments will be needed to clarify how essential CTCF and cohesin are in determining the 3D landscape in ESCs.

4 Genome Organization During Linage Commitment from ESCs

ESCs have the potential to derive all cell linages that form the embryo proper. In order to describe how genome higher-order chromatin architecture changes upon early linage commitment, Dixon et al. [52], performed HiC using four different cell lineages derived from human ESCs. During differentiation, TADs seem to be conserved between ESCs and the differentiated tissues although there is prominent rewiring of interactions within TADs specially from H3K4me1 marked regions. Interestingly, compartments have a major reorganization during the differentiation process. In particular the B compartment seems to expand in committed cells versus ESCs, which is concordant with heterochromatin expansion in differentiated cells [52].

Promoter Capture Hi-C analysis during human ESC differentiation to NEC (neuroectodermal cells) has shown extensive dynamics in the connectivity of gene promoters upon differentiation as well as prominent changes in the chromatin state of the cis-regulatory units identified as interacting with target promoters [53].

5 Genome Topology and Early Embryogenesis In Vivo

During embryonic development, the spatial and temporal control of gene expression is crucial to produce a viable organism and chromatin architecture is fundamental in regulating such process [54–57]. Single cell HiC analysis of the oocyte to zygote transition in mice has revealed unique features of the totipotent genome architecture. In particular, the maternal chromatin seems to organize in TADs and loops but does not distributes into compartments as other genomes during interphase [19]. Further single cell studies will be needed to fully catalogue the epigenetic and topological properties of the totipotent genome.

Outstanding work performed in Drosophila zygotes has shown that the spatial organization of the genome is established during a small period that coincides with its transcriptional activation. Interestingly, the appearance of TAD boundaries is independent of transcriptional activity and involves the transcription factor Zelda. Thus, it seems that 3D organization comes first than transcription in the very first genomes of the Drosophila embryo [58].

Distal regulatory elements such as enhancers have a key role in gene expression control, and the tridimensional organization of the genome provides a framework to generate relevant structural and functional contacts between regulatory elements and their target genes [9, 59]. In Drosophila long-range interactions have been identified between genes and multiple enhancers during embryogenesis. These long-range interactions contribute to control the transcriptional output of specific genes such as *Brinker*, *Shavenbaby*, and the *Hox* genes [9, 60, 61]. Enhancers can have redundant activity (named shadow enhancers) such that the alteration of one does not necessarily causes aberrant phenotypes providing robustness to the regulation system to ensure proper development [60, 62].

In the developing mesoderm, there are 1055 shadow enhancers reported, associated to 319 genes of which 40% have two shadow enhancers and the rest have three to eight shadow enhancers [60]. Using 4C–seq, 92 enhancers where chosen as viewpoints to generate high-resolution interaction maps of mesodermal cells and whole embryos in two developmental stages, during which the embryo undergoes robust changes in morphology and gene transcription [9]. The high-resolution maps show 1036 unique interactions, each enhancer interacting with ten genomic regions on average. Most interactions are maintained independently of transcription activity, in mesodermal cells and whole embryos in the two evaluated stages [9]. The preformed interactions found between enhancers and promoters in the transcriptionally off state, present accumulated levels of paused RNA pol II [55]. Stable interactions (independent on transcriptional state) have also been described in human primary IMR90 fibroblasts, between enhancers and promoters [63]. How dynamic chromatin loops and larger topological structures are remains a very active research area in the field.

5.1 Topology Modulates Hox Gene Expression

The Homeotic gene clusters (Hox) orchestrate anterior–posterior body axis and are key for correct embryo development and patterning in bilaterian animals. In Drosophila two clusters Antennapedia (ANT-C) and Bithorax (BX-C) are organized in genomic loci containing six genes (*Lab*, *Pd*, *Dfd*, *Scr*, *Ftz*, and *Antp*) and three genes (*Ubx*, *Abd-A*, and *Abd-B*) respectively. Upstream of the *Scr* gene is the *Ftz* gene, and between them there are enhancer elements that control the expression of both at different developmental stages [64].

In early development, *Scr* is inactive while *Ftz* is actively transcribed. A chromatin loop brings the enhancer and the *Ftz* gene together enriching the gene of permissive histone modifications and allowing its transcription. Outside of the loop the chromatin presents repressive histone marks. At late development this loop is disrupted and the *Scr* loci gains active chromatin marks and now the enhancers act on that promoter [64]. This is an example of dynamic contacts where the interactions are formed to activate transcription, in contrast to what was previously described. SF1 is the factor orchestrating the dynamic chromatin looping in early and late development in the case of the *Ftz* gene loop, while insulator-structural proteins maintain the TAD structure [64].

In mammals there are 39 *Hox* genes distributed in four clusters (*HoxA*, *HoxB*, *HoxC*, and *HoxD*) located in different chromosomes [57, 65]. Based on homology and position, the *Hox* genes are classified in 13 paralogous groups, where the Hox1 is located at the 3′ and Hox13 is at the 5′ side of the corresponding cluster [66].

HoxA and *HoxD* gene expression patterns are similar and both are critical for limb bud development. *HoxA* genes also play a role in the proper function of neural crest cells [67]. The *HoxD* cluster is transcribed following a clear collinear strategy. In an early phase of forelimb bud development, *Hoxd1-Hoxd10* genes are active and enriched in histone marks H3K4me3 and H3K27ac, whereas *Hoxd11-Hoxd13* genes are practically inactive and enriched in H3K27me3 mark associated with Polycomb and this is inversed later in development. The HoxD cluster resides at the boundary between two TADs termed telomeric and centromeric domains, both matching with gene deserts containing enhancers; at least six enhancers in the centromeric and two in the telomeric domains (Fig. 2). The regulatory elements in the telomeric TAD are also marked with H3K27ac and contact the *HoxD* cluster in early limb development to form the arms and forearms or legs and lower legs (Fig. 2, top). Later in development the telomeric domain is enriched in the repressive Polycomb mark H3K27me3 and now

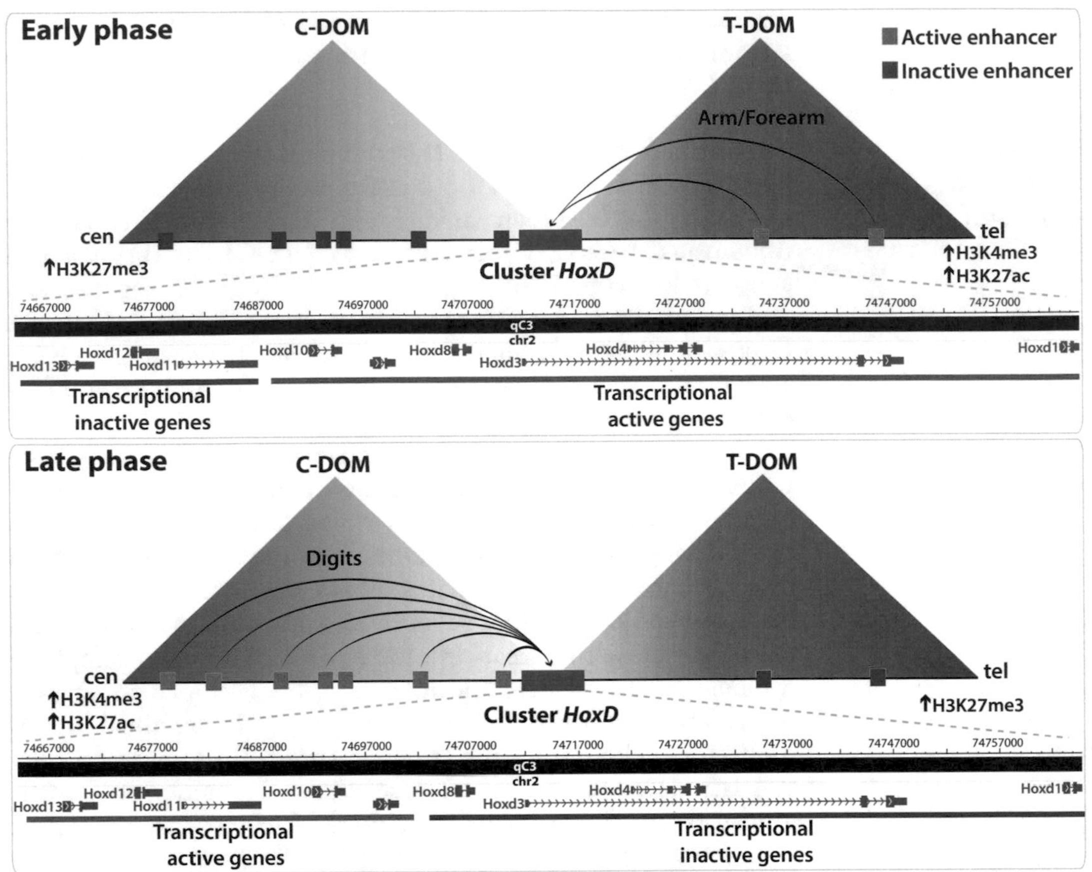

Fig. 2 HoxD genes regulation during limb development. The *HoxD* cluster is located at a boundary between C-DOM (centromeric TAD) and T-DOM (telomeric TAD) domains. During early arm and forearm development, the T-DOM is occupied by active histone marks (H3K4me3 and H3K27ac) and the two enhancers within this domain act on *Hoxd1-Hoxd10* genes and promote their expression. Meanwhile, the C-DOM is enriched in the repressive histone mark H3K27me3. In late development, a switch in gene expression and histone marks occurs, the enhancers within the C-DOM now promote the expression of the *Hoxd9-Hoxd13* genes and digits are formed [67]

distal regulatory elements within the centromeric TAD, enriched in H3K27ac contact the *HoxD* genes to form digits (Fig. 2, bottom). This represents an outstanding example of a structural regulatory chromatin landscape modulating long-range contacts involved in determining the temporal and spatial transcription rate of the *Hox* genes during limb development [67]. It is interesting to emphasize that the *HoxD* gene cluster is located at the TAD boundary and is the shift of interactions swinging from one TAD to the other, which confers the transcriptional regulation. Topological regulation of gene expression by switching TADs from the boundaries might be more common than previously thought.

CTCF insulates these adjacent yet antagonistic chromatin domains. Deletion of CTCF binding sites within the *HoxD* cluster results in the expansion of active chromatin into the repressive domain [68–70]. Although the precise mechanism that promotes the switch is unknown, HOXA13 protein seems to play an important role in the process [68]. After the early phase of limb development HOXA13 binds the HoxD cluster and activates transcription in the centromeric domain. At the same time, it has a repressive effect on the telomeric domain at the chromatin level, and this effect is enhanced by HOXD13 protein. The transcriptional inactivation of *Hoxa13* and *Hoxd13* genes results in the lack of digits [68].

6 TADs Disruption and Disease

Recent evidence has shown the key regulatory role of the topological partition of the genome into TADs. Alterations in TADs structure can conduce to deregulation of long-range interactions producing aberrant gene expression patterns and severe phenotypes, as exemplified below.

6.1 Structural Variations at the Epha4 Gene Locus

The *EPHA4* gene encodes the Ephrin receptor A4, involved in axon growth guidance, formation of boundaries between tissues, cell migration and segmentation. The gene resides in a 2 Mb gene desert TAD, flanked by two domains that contain the *WNT6* and *Ihh* genes within the centromeric TAD, and *PAX3* gene inside the telomeric TAD. Structural variants in *EPHA4* locus are associated with rare limb malformations: brachydactyly, F-syndrome/syndactyly, and polydactyly. These phenotypes can be reproduced by genetic engineering through CRISPR/Cas9 in mouse [71].

In wild type embryos *Epha4* gene expression is regulated by an enhancer located inside the same TAD. Genes located within the centromeric TAD (*Wnt6 and Ihh*) are transcriptionally inactive. The F-syndrome (fused fingers) is caused by an inversion or duplication between the TAD containing the *Epha4* gene and the centromeric TAD containing *Wnt6* and *Ihh genes.* The inverted region contains the enhancer and the *Ihh* gene. When the inversion is present, the enhancer is relocated to the centromeric TAD resulting in *Wnt6* gene expression activation (Fig. 3, middle). Duplication between *Ihh* gene and the most centromeric part of the *Epha4* TAD results in massive polydactyly with incorrect activation of *Ihh* by the enhancer (Fig. 3, top) [71]. Brachydactyly (short digits) results from the deletion of a region containing the *Epha4* gene and its TAD boundary. In this topological landscape, now the enhancer is able to incorrectly activate *Pax3* on the telomeric TAD (Fig. 3, bottom). In summary, depending on where the breakpoint is located and if it alters the location of the TAD boundaries, the

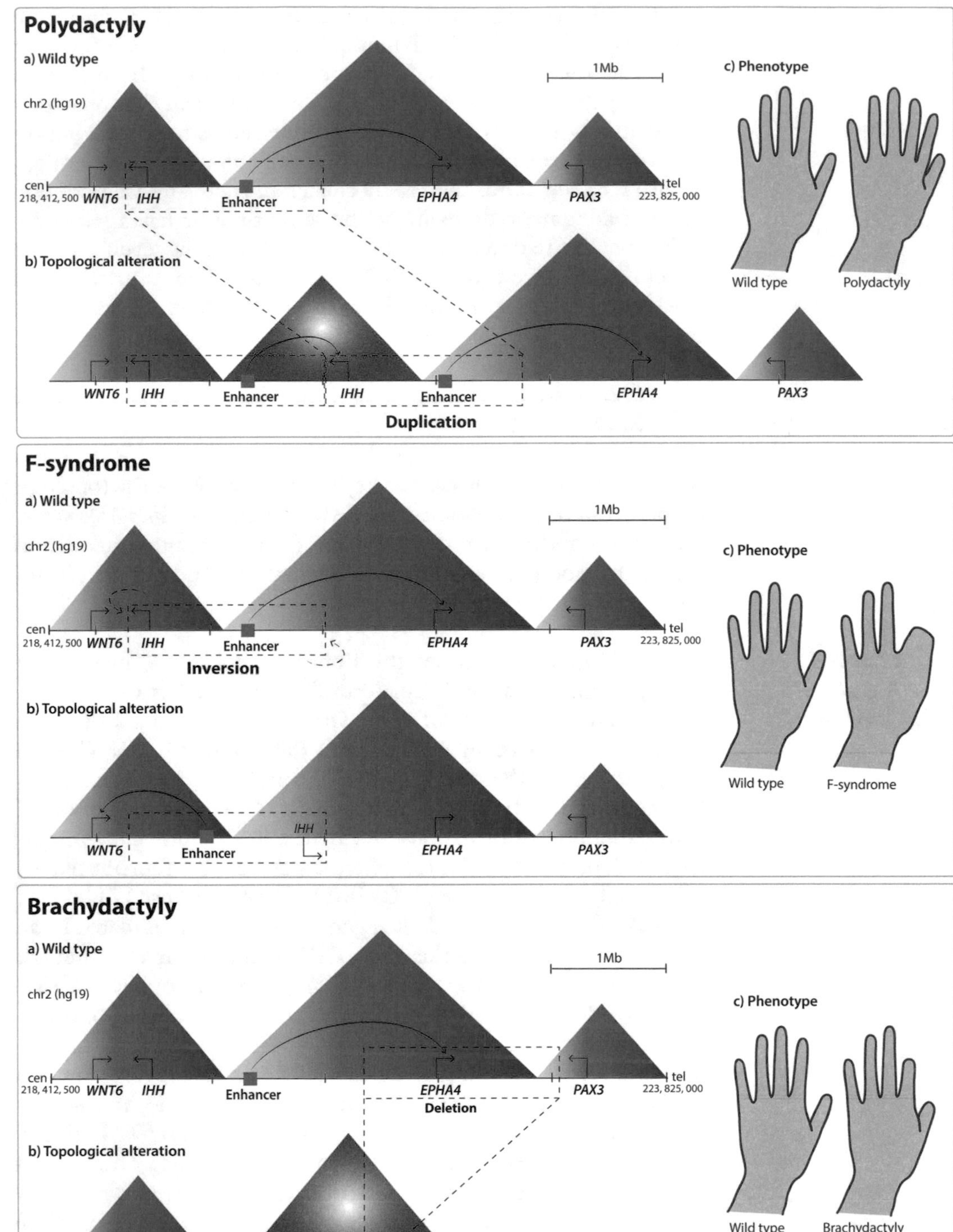

Fig. 3 Topological structure alterations in limb development. The *EPHA4* gene and enhancer are located in a large gene deserted TAD. In wild type embryos the enhancer promotes *EPHA4* expression leading to proper digit development; this TAD is bordered by a centromeric (cen) and a telomeric (tel) TADs, whose genes are transcriptionally inactive.

Epha4 enhancer is able to activate genes (*Pax3* in brachydactyly, *Wnt6* in F-syndrome and *Ihh* in polydactyly syndrome) located in other TADs resulting in severe malformations.

6.2 Liebenberg Syndrome

The Liebenberg syndrome is an autosomal-dominant alteration wherein the arms acquire leg characteristics. This condition is caused by deletion of a region including the boundary of the TAD containing the *H2AFY* gene (Fig. 4, top). Upstream of *H2AFY* there is an enhancer and downstream, the *PITX1* gene is located in the next TAD. When the deletion is present the enhancer incorrectly activates *PITX1* transcription leading to improper anterior limb formation [72, 73].

6.3 Mesomelic Dysplasia

The mesomelic dysplasia refers to a skeletal disorder where the middle limb segments are shortened. There are 11 types of mesomelia; one of them is the Savarirayan type, in which the phenotype displays shortness of the lower limb middle segment. In patients with this disorder, a deleted region was identified comprising the genes *MBOAT1*, *E2F3*, *CDKAL1*, and *SOX4*. *MBOAT1* gene is located at a TAD boundary. Inside the TAD upstream *MBOAT1*, the *E2F3* and *CDKAL1* genes are located. *SOX4* and two enhancers reside in the following topological domain upstream (Fig. 4, middle). Downstream *MBOAT1*, the *ID4* gen is located. In the presence of the genetic deletion the TAD boundaries are lost and one big TAD is formed instead of the three original ones. Within this domain now, incorrect interactions between the enhancers and *ID4* gene promoter are formed causing its misexpression (Fig. 4, middle) [74].

6.4 Autosomal-Dominant Adult-Onset Demyelinating Leukodystrophy (ADLD)

The autosomal dominant adult onset demyelinating leukodystrophy (ADLD) is a neurological disorder wherein *LAMIN B1* overexpression and protein accumulation generates central nervous system demyelination. Deletion of the TAD boundary upstream of the *LMNB1* gene, results in TAD fusion and ectopic interactions between enhancers and the *LMNB1* gene promoter (Fig. 4, bottom) [73, 75].

Fig. 3 (continued) Chromosome conformation alterations in this region cause digits malformations. Duplication of a region that includes *IHH* gen and enhancer (dash rectangle) resulting in a new TAD (orange TAD) and *IHH* gene expression leads to polydactyly. Phenotypes for human and mouse have been described in [71, 76]. The F-syndrome is caused by an inversion resulting in *WNT6* gene expression. The phenotype exhibits finger fusion [71, 77]. A genomic deletion resulting in TAD boundary and *EPHA4* gene deletion leads to the formation of a new TAD (orange TAD) and incorrect activation of *PAX3* by the enhancer. This genomic and topological alteration produces short digits (brachydactyly) [71, 78]

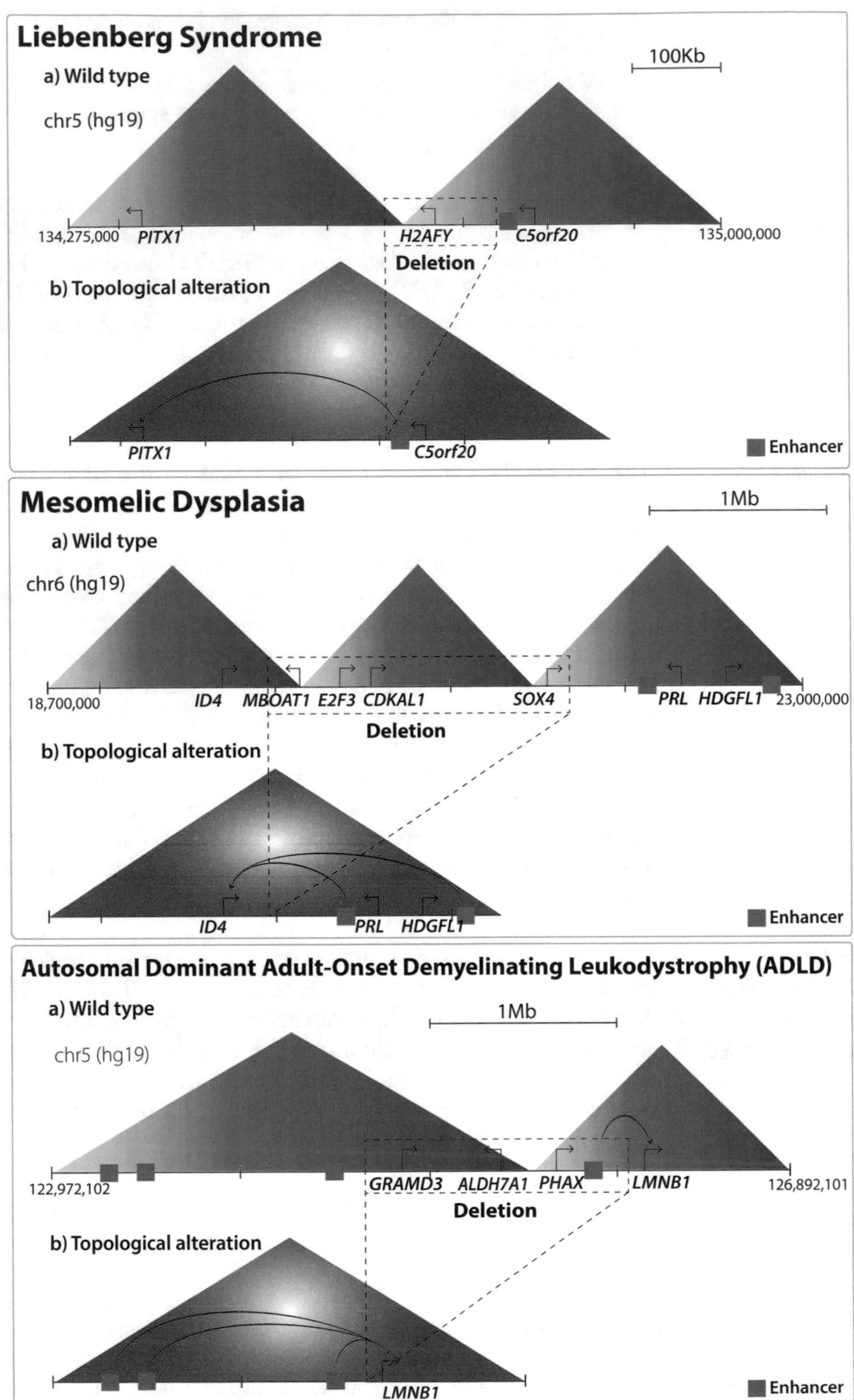

Fig. 4 Other topological alterations. Liebenberg syndrome is produced by a TAD boundary and *H2AFY* gen deletion (dashed rectangle). In this context the enhancer upstream of *H2AFY* is now able to incorrectly activate *PITX1* gene transcription (for phenotype *see* [72, 79]). Mesomelic dysplasia implies a genomic deletion covering

7 Concluding Remarks

3D genome organization is a key regulator of genomic function. There has been remarkable progress describing chromatin structure at high resolutions and teasing apart its hierarchical nature. From chromosome territories to chromatin loops connecting genes and regulatory elements and other unknown sequences in the genome, we have now the possibility to interfere with such structures and measure its consequences at different functional levels including transcription, replication, recombination and repair. Furthermore, we can continue to add molecular players participating in chromatin conformation to integrate the relative location of chromatin within the nucleus with its fine 3D structure.

The relevance of genome topology is emphasized in the early embryo and ESCs in which transcription factors and protein complexes that maintain the pluripotent state and specify linage commitment, operate through modulation of 3D genome structure. The importance of chromatin organization is also highlighted by analysis of structural variations affecting topological genomic attributes that result in severe phenotypes during embryogenesis and development. Finally, genome editing can now be used systematically to alter structural genomic features, relocate chromatin segments toward different nuclear compartments and measure its functional consequences.

Acknowledgments

This work was funded by the Technology Innovation and Research Support Programme (PAPIIT) num. IA201817.

Fig. 4 (continued) *MBOAT1, E2F3, CDKAL1*, and *SOX4* genes (dashed rectangle). TAD boundaries are removed and only one TAD is formed (orange TAD) allowing *ID4* transcription by promiscuous interactions with enhancers leading to a shortened lower limb middle segment (for phenotype *see* [74]). ADLD is caused by deletion of several genes and a TAD boundary (dashed rectangle), resulting in ectopic interactions between the enhancers and LMNB1 (for phenotype *see* [73, 75])

References

1. Jost KL, Bertulat B, Cardoso MC (2012) Heterochromatin and gene positioning: Inside, outside, any side? Chromosoma 121:555–563. https://doi.org/10.1007/s00412-012-0389-2
2. Denker A, De Laat W (2016) The second decade of 3C technologies: detailed insights into nuclear organization. Genes Dev 30: 1357–1382. https://doi.org/10.1101/gad.281964.116
3. Bolzer A, Kreth G, Solovei I et al (2005) Three-dimensional maps of all chromosomes in human male fibroblast nuclei and prometaphase rosettes. PLoS Biol 3:0826–0842. https://doi.org/10.1371/journal.pbio.0030157
4. Fraser J, Ferrai C, Chiariello AM et al (2015) Hierarchical folding and reorganization of chromosomes are linked to transcriptional changes in cellular differentiation. Mol Syst Biol 11:1–14. https://doi.org/10.15252/msb
5. Dekker J (2006) The three "C" s of chromosome conformation capture: controls, controls, controls. Nat Methods 3:17–21. https://doi.org/10.1038/NMETH823
6. Brant L, Georgomanolis T, Nikolic M et al (2016) Exploiting native forces to capture chromosome conformation in mammalian cell nuclei. Mol Syst Biol 561:1–23. https://doi.org/10.15252/msb.20167311
7. Simonis M, Klous P, Splinter E et al (2006) Nuclear organization of active and inactive chromatin domains uncovered by chromosome conformation capture-on-chip (4C). Nat Genet 38:1348–1354. https://doi.org/10.1038/ng1896
8. Schwartzman O, Mukamel Z, Oded-Elkayam N et al (2016) UMI-4C for quantitative and targeted chromosomal contact profiling. Nat Methods 13:685–691. https://doi.org/10.1038/nmeth.3922
9. Ghavi-Helm Y, Klein FA, Pakozdi T et al (2014) Enhancer loops appear stable during development and are associated with paused polymerase. Nature 512:96–100. https://doi.org/10.1038/nature13417
10. Nora EP, Lajoie BR, Schulz EG et al (2012) Spatial partitioning of the regulatory landscape of the X-inactivation centre. Nature 485:381–385. https://doi.org/10.1038/nature11049
11. Dostie J, Richmond TA, Arnaout RA et al (2006) Chromosome conformation capture carbon copy (5C): a massively parallel solution for mapping interactions between genomic elements. Genome Res 16:1299–1309. https://doi.org/10.1101/gr.5571506
12. Lieberman-Aiden E, vand Berkum N (2009) Comprehensive mapping of long range interactions reveals folding principles of the human genome. Science 326:289–293. https://doi.org/10.1126/science.1181369.Comprehensive
13. Hughes JR, Roberts N, McGowan S et al (2014) Analysis of hundreds of cis-regulatory landscapes at high resolution in a single, high-throughput experiment. Nat Genet 46:205–212. https://doi.org/10.1038/ng.2871
14. Schoenfelder S, Sugar R, Dimond A et al (2015) Polycomb repressive complex PRC1 spatially constrains the mouse embryonic stem cell genome. Nat Genet 47:1179–1186. https://doi.org/10.1038/ng.3393
15. Mifsud B, Tavares-Cadete F, Young AN et al (2015) Sup mapping long-range promoter contacts in human cells with high-resolution capture Hi-C. Nat Genet 47:598–606. https://doi.org/10.1038/ng.3286
16. Nagano T, Lubling Y, Stevens TJ et al (2013) Single-cell Hi-C reveals cell-to-cell variability in chromosome structure. Nature 502:59–64. https://doi.org/10.1038/nature12593
17. Nagano T, Lubling Y, Varnai C et al (2017) Cell-cycle dynamics of chromosomal organization at single-cell resolution. Nature 547: 61–67. https://doi.org/10.1038/nature23001
18. Stevens TJ, Lando D, Basu S et al (2017) 3D structures of individual mammalian genomes studied by single-cell Hi-C. Nature 544:59–64. https://doi.org/10.1038/nature21429
19. Flyamer IM, Gassler J, Imakaev M et al (2017) Single-cell Hi-C reveals unique chromatin reorganization at oocyte-tozygote transition. Nat Publ Gr 544:1–17. https://doi.org/10.1038/nature21711
20. Fullwood MJ, Liu MH, Pan YF et al (2009) An oestrogen-receptor-alpha-bound human chromatin interactome. Nature 462:58–64. https://doi.org/10.1038/nature08497
21. Mumbach MR, Rubin AJ, Flynn RA et al (2016) HiChIP: efficient and sensitive analysis of protein-directed genome architecture. bioRxiv: 73619. https://doi.org/10.1101/073619
22. Dixon JR, Selvaraj S, Yue F et al (2012) Topological domains in mammalian genomes identified by analysis of chromatin interactions. Nature 485:376–380. https://doi.org/10.1038/nature11082
23. Nora EP, Lajoie BR, Schulz EG et al (2013) Spatial partitioning of the regulatory landscape of the X- inactivation center. Nature 485: 381–385. https://doi.org/10.1038/nature 11049.Spatial

24. Sexton T, Yaffe E, Kenigsberg E et al (2012) Three-dimensional folding and functional organization principles of the Drosophila genome. Cell 148:458–472. https://doi.org/10.1016/j.cell.2012.01.010
25. Mizuguchi T, Fudenberg G, Mehta S et al (2014) Cohesin-dependent globules and heterochromatin shape 3D genome architecture in S. pombe. Nature 516:432–435. https://doi.org/10.1038/nature13833
26. Crane E, Bian Q, Mccord RP et al (2015) Condensin-driven remodelling of X chromosome topology during dosage compensation. Nature. https://doi.org/10.1038/nature14450
27. Le TBK, Imakaev MV, Mirny LA, Laub MT (2013) High-resolution mapping of the spatial organization of a bacterial chromosome. Science 342:731–735
28. Rao SSP, Huntley MH, Durand NC et al (2014) A 3D map of the human genome at kilobase resolution reveals principles of chromatin looping. Cell 159:1665–1680. https://doi.org/10.1016/j.cell.2014.11.021
29. Nora EP, Goloborodko A, Valton AL et al (2017) Targeted degradation of CTCF decouples local insulation of chromosome domains from higher-order genomic compartmentalization. Cell 169:930-944e.22. http://dx.doi.org/10.1016/j.cell.2017.05.004
30. Eagen K, Lieberman Alden E, Kornberg DR (2017) Polycomb-mediated chromatin loops revealed by a sub-kilobase resolution chromatin interaction map. bioRxiv. https://doi.org/10.1101/099804
31. Melcer S, Meshorer E (2010) Chromatin plasticity and genome organization in pluripotent embryonic stem cells. Curr Opin Cell Biol 22:334–341. https://doi.org/10.1016/j.ceb.2010.02.001
32. Meshorer E, Yellajoshula D, George E et al (2006) Hyperdynamic plasticity of chromatin proteins in pluripotent embryonic stem cells. Dev Cell 10:105–116. https://doi.org/10.1016/j.devcel.2005.10.017
33. Gaspar-Maia A, Alajem A, Meshorer E, Ramalho-Santos M (2011) Open chromatin in pluripotency and reprogramming. Nat Rev Mol Cell Biol 12:36–47. https://doi.org/10.1038/nrm3036
34. Zhu J, Adli M, Zou JY et al (2013) Genome-wide chromatin state transitions associated with developmental and environmental cues. Cell 152:642–654. https://doi.org/10.1016/j.cell.2012.12.033
35. Ficz G, Heintzmann R, Arndt-Jovin DJ (2005) Polycomb group protein complexes exchange rapidly in living Drosophila. Development 132:3963–3976. https://doi.org/10.1242/dev.01950
36. Ren X, Vincenz C, Kerppola TK (2008) Changes in the distributions and dynamics of polycomb repressive complexes during embryonic stem cell differentiation. Mol Cell Biol 28:2884–2895. https://doi.org/10.1128/MCB.00949-07
37. Denholtz M, Bonora G, Chronis C et al (2013) Long-range chromatin contacts in embryonic stem cells reveal a role for pluripotency factors and polycomb proteins in genome organization. Cell Stem Cell 13:602–616. https://doi.org/10.1016/j.stem.2013.08.013
38. Kundu S, Ji F, Sunwoo H et al (2017) Polycomb Repressive Complex 1 Generates Discrete Compacted Domains that Change during Differentiation. Mol Cell 65:432–445.e6. https://doi.org/10.1016/j.molcel.2017.01.009
39. Richly H, Aloia L, Di Croce L (2011) Roles of the polycomb group proteins in stem cells and cancer. Cell Death Dis 2:e204. https://doi.org/10.1038/cddis.2011.84
40. Wang W, Quin J-J, Voruganti S et al (2015) Polycomb Group (PcG) proteins and human cancers: multifaceted functions and therapeutic implications. Med Res Rev 22:134–139. https://doi.org/10.1177/0963721412473755. Surging
41. Yamanaka S, Blau HM (2010) Nuclear reprogramming to a pluripotent state by three approaches. Nature 465:704–712. https://doi.org/10.1038/nature09229
42. Whyte WA, Orlando DA, Hnisz D et al (2013) Master transcription factors and mediator establish super-enhancers at key cell identity genes. Cell 153:307–319. https://doi.org/10.1016/j.cell.2013.03.035
43. Li Y, Rivera CM, Ishii H et al (2014) CRISPR reveals a distal super-enhancer required for Sox2 expression in mouse embryonic stem cells. PLoS One 9:1–17. https://doi.org/10.1371/journal.pone.0114485
44. de Wit E, Bouwman BA, Zhu Y et al (2013) The pluripotent genome in three dimensions is shaped around pluripotency factors. Nature 501:227–231. https://doi.org/10.1038/nature12420
45. Wei Z, Gao F, Kim S et al (2013) Klf4 organizes long-range chromosomal interactions with the OCT4 locus inreprogramming and-pluripotency. Cell Stem Cell 13:36–47. https://doi.org/10.1016/j.stem.2013.05.010
46. Monahan K, Rudnick ND, Kehayova PD et al (2012) Role of CCCTC binding factor (CTCF)

and cohesin in the generation of single-cell diversity of protocadherin-α gene expression. Proc Natl Acad Sci U S A 109:9125–9130. https://doi.org/10.1073/pnas.1205074109

47. Zuin J, Dixon JR, van der Reijden MIJA et al (2014) Cohesin and CTCF differentially affect chromatin architecture and gene expression in human cells. Proc Natl Acad Sci U S A 111:996–1001. https://doi.org/10.1073/pnas.1317788111
48. Merkenschlager M, Nora EP (2016) CTCF and cohesin in genome folding and transcriptional gene regulation. Annu Rev Genomics Hum Genet 17:17–43. https://doi.org/10.1146/annurev-genom-083115-022339
49. Nitzsche A, Paszkowski-Rogacz M, Matarese F et al (2011) RAD21 cooperates with pluripotency transcription factors in the maintenance of embryonic stem cell identity. PLoS One. https://doi.org/10.1371/journal.pone.0019470
50. Dowen JM, Fan ZP, Hnisz D et al (2014) Control of cell identity genes occurs in insulated neighborhoods in mammalian chromosomes. Cell 159:374–387. https://doi.org/10.1016/j.cell.2014.09.030
51. Kubo N, Ishii H, Gorkin D et al (2017) Preservation of chromatin organization after acute loss of CTCF in mouse embryonic stem cells 2 3. bioRxiv. https://doi.org/10.1101/118737
52. Dixon JR, Jung I, Selvaraj S et al (2015) Chromatin architecture reorganization during stem cell differentiation. Nature 518:331–336. https://doi.org/10.1038/nature14222
53. Freire-pritchett P, Schoenfelder S, Várnai C, Steven W (2017) Global reorganisation of cis-regulatory units upon lineage commitment of human embryonic stem cells. elife 6:pii:e21926. https://doi.org/10.7554/eLife.21926
54. de Laat W, Duboule D (2013) Topology of mammalian developmental enhancers and their regulatory landscapes. Nature 502:499–506. https://doi.org/10.1038/nature12753
55. Mallo M, Alonso CR (2013) The regulation of Hox gene expression during animal development. Development 140:3951–3963. https://doi.org/10.1242/dev.068346
56. Montavon T, Duboule D (2013) Chromatin organization and global regulation of Hox gene clusters. Philos Trans R Soc Lond Ser B Biol Sci 368:20120367. https://doi.org/10.1098/rstb.2012.0367
57. Montavon T, Duboule D (2012) Landscapes and archipelagos: spatial organization of gene regulation in vertebrates. Trends Cell Biol 22:347–354. https://doi.org/10.1016/j.tcb.2012.04.003
58. Hug CB, Grimaldi AG, Kruse K, Vaquerizas JM (2017) Chromatin architecture emerges during zygotic genome activation independent of transcription. Cell 169:216–228.e19. https://doi.org/10.1016/j.cell.2017.03.024
59. Ma Z, Li M, Roy S et al (2016) Chromatin boundary elements organize genomic architecture and developmental gene regulation in Drosophila Hox clusters. World J Biol Chem 7:223–230. https://doi.org/10.4331/wjbc.v7.i3.223
60. Cannavò E, Khoueiry P, Garfield DA et al (2016) Shadow enhancers are pervasive features of developmental regulatory networks. Curr Biol 26:38–51. https://doi.org/10.1016/j.cub.2015.11.034
61. Perry MW, Boettiger AN, Levine M (2011) Multiple enhancers ensure precision of gap gene-expression patterns in the Drosophila embryo. PNAS 108:1–12. https://doi.org/10.1073/pnas.1109873108
62. Hong J-W, Hendrix DA, Levine MS (2008) Shadow enhancers as a source of evolutionary novelty. Science 321:1314. https://doi.org/10.1126/science.1160631
63. Jin F, Li Y, Dixon JR et al (2013) A high-resolution map of the three-dimensional chromatin interactome in human cells. Nature 503:290–294. https://doi.org/10.1038/nature12644
64. Li M, Ma Z, Liu JK et al (2015) An organizational hub of developmentally regulated chromatin loops in the drosophila antennapedia complex. Mol Cell Biol 35:MCB.00663-15. https://doi.org/10.1128/MCB.00663-15. Address
65. Pindyurin AV, van Steensel B (2012) Hox in space. Nucleus 3:118–122. https://doi.org/10.4161/nucl.19159
66. Montavon T, Soshnikova N (2014) Hox gene regulation and timing in embryogenesis. Semin Cell Dev Biol 34:76–84. https://doi.org/10.1016/j.semcdb.2014.06.005
67. Andrey G, Montavon T, Mascrez B et al (2012) A switch between topological domains underlies HoxD genes collinearity in mouse limbs. Nat Rev Genet 13:613–626. https://doi.org/10.1038/nrg3207
68. Beccari L, Yakushiji-Kaminatsui N, Woltering JM et al (2016) A role for HOX13 proteins in the regulatory switch between TADs at the HoxD locus. Genes Dev 30:1172–1186. https://doi.org/10.1101/gad.281055.116
69. Soshnikova N, Montavon T, Leleu M et al (2010) Functional analysis of CTCF during mammalian limb development. Dev Cell 19:819–830. https://doi.org/10.1016/j.devcel.2010.11.009

70. Narendra V, Rocha PP, An D et al (2015) Transcription. CTCF establishes discrete functional chromatin domains at the Hox clusters during differentiation. Science 347: 1017–1021. https://doi.org/10.1126/science.1262088
71. Lupiáñez DG, Kraft K, Heinrich V et al (2015) Disruptions of topological chromatin domains cause pathogenic rewiring of gene-enhancer interactions. Cell 161:1012–1025. https://doi.org/10.1016/j.cell.2015.04.004
72. Spielmann M, Brancati F, Krawitz PM et al (2012) Homeotic arm-to-leg transformation associated with genomic rearrangements at the PITX1 locus. Am J Hum Genet 91:629–635. https://doi.org/10.1016/j.ajhg.2012.08.014
73. Lupiáñez DG, Spielmann M, Mundlos S (2016) Breaking TADs: how alterations of chromatin domains result in disease. Trends Genet 32:225–237. https://doi.org/10.1016/j.tig.2016.01.003
74. Flottmann R, Wagner J, Kobus K et al (2015) Microdeletions on 6p22.3 are associated with mesomelic dysplasia Savarirayan type. J Med Genet 52:476–483. https://doi.org/10.1136/jmedgenet-2015-103108
75. Giorgio E, Robyr D, Spielmann M et al (2014) A large genomic deletion leads to enhancer adoption by the lamin B1 gene: a second path to autosomal dominant adult-onset demyelinating leukodystrophy (ADLD). Hum Mol Genet 24:3143–3154. https://doi.org/10.1093/hmg/ddv065
76. Chakraborty PB, Marjit B, Dutta S, De A (2007) Polydactyly: a case study. J Anat Soc India 56:35–38
77. Flatt AE (2005) Webbed fingers. Proc (Bayl Univ Med Cent) 18:26–37
78. Temtamy SA, Aglan MS (2008) Brachydactyly. Orphanet J Rare Dis 3:15. https://doi.org/10.1186/1750-1172-3-15
79. Mennen U, Mundlos S, Spielmann M (2014) The Liebenberg syndrome: in depth analysis of the original family. J Hand Surg 39:919–925. https://doi.org/10.1177/1753193413502162

Check for updates

Chapter 16

Genome Editing During Development Using the CRISPR-Cas Technology

Rodrigo G. Arzate-Mejía, Paula Licona-Limón, and Félix Recillas-Targa

Abstract

Over the years, the study of gene function during development involved the implementation of sophisticated transgenic strategies to visualize how organisms change during their lifetime. These strategies are diverse and extremely useful and allowed the discovery of some of the fundamental mechanisms governing organism's development. Such strategies can be time-consuming, in some cases expensive, and require complex infrastructure. With the advent of the genome editing CRISPR-Cas9 RNA-guided DNA endonuclease system a tremendous progress has been achieved in manipulating diverse organisms and cell types. In recent years this system has contributed importantly to the design of novel experimental strategies to further understand developmental processes, to generate genetically modified animal models, and develop disease models. Here we highlight examples in which the genome editing CRISPR-Cas9 system has been employed to understand the mechanisms controlling embryonic development and disease.

Key words Genome editing, ZFN, TALEN, CRISPR-Cas6, Development, Disease, DNA repair, iPSC, ESC, Chromatin

1 Introduction

Since early 1970s several research groups realized the usefulness of the genetic manipulation of live organisms in particular to understand the mechanisms governing animal development. In 1974, Rudolf Jaenish and Beatrice Mintz generated the first transgenic mouse taking advantage of previous advances of many scientists including Stanley Cohen and Herbert Boyer that allowed the cloning of DNA fragments into plasmid vectors for the subsequent generation of transgenic animals [1, 2]. However, these approaches have disadvantages including low reproducibility of phenotypes due to variability of transgene expression owing to chromosomal position effects [3]. In certain contexts, the copy number of the integrated construct may also distort the expected pattern of transgene expression. Twenty years after generation of

Paul Delgado-Olguin (ed.), *Mouse Embryogenesis: Methods and Protocols*, Methods in Molecular Biology, vol. 1752, https://doi.org/10.1007/978-1-4939-7714-7_16,

the first transgenic mouse, alternatives to directly manipulate specific genomic regions were explored.

One of the first innovations emerged with the zinc-finger DNA-binding proteins (ZFPs) as a tool to direct the activity of endonucleases to specific nucleotide triplets [4]. This strategy evolved with the incorporation of the *FokI* endonuclease to the ZFP [5]. This zinc-finger protein-based strategy was used for the first time in 2009 to generate knockout rats by microinjection of embryos with zinc-finger nucleases against the rat *immunoglobulin M* and *Rad38* genes [6]. Today it is known that the zinc-finger nucleases are a successful approach for gene targeting, but its use is limited by unspecific interactions with DNA sequences.

To further increase the DNA binding specificity, Transcription-Activator-Like Effect Nucleases or TALENs were developed 2 years later. The TALEN method was nominated as method of the year in 2011 [7]. TALEN proteins possess a domain for DNA binding that consists of monomers with the capacity to recognize specifically one nucleotide. Two amino acid residues in the monomer are responsible for recognition and binding. Then, the *FokI* catalytic domain was fused allowing targeting of the TALEN proteins and endonuclease activity at desired sequences with higher specificity than ZFNs [7, 8]. TALENs rapidly became the most popular method for genome editing. Unfortunately, both systems the ZFNs and TALENs, revealed to be time consuming and costly. In addition, the high content of repetitive sequences in TALEN vectors can induce a high degree of instability in the constructs.

2 The Genome Editing CRISPR-Cas9 System

A significant step forward came with the advent of the CRISPR-Cas9 system for genome editing and the demonstration of efficient genome modification in zebrafish, mice, and mammalian cells [9, 10]. The CRISPR-Cas system stands for Clustered Regularly Interspaced Short Palindromic Repeat (CRISPR) sequences, which in conjunction with the CRISPR-associated (Cas) nuclease, constitute the CRISPR-Cas adaptative immune system in prokaryotes [10]. The initial discoveries related to the CRISPR system came in late 1980s, and more than 20 years of intense research by many groups culminated in two seminal reports. The laboratories of Jennifer Doudna and Emmanuelle Charpentier described the reconstitution in vitro of the CRISPR-Cas9 system and demonstrated its capacity to generate double strand breaks (DSBs) and the integration of homologous sequences to the CRISPR RNA (crRNA) spacer region [11, 12]. The CRISPR system includes an RNA-guided endonuclease protein from *Streptococcus pyogenes* known as Cas9 that uses two small RNAs, a CRISPR-associated

RNA (crRNA), and the *trans*-acting crRNA (tracRNA) for sequence-specific DNA cleavage [11, 13]. Further studies allowed the design of a chimeric single guide RNA (sgRNA) located between the crRNA and tracRNA that recapitulates the structure and function of the crRNA-tracRNA complex, simplifying the system and allowing efficient and specific double-strand breaks in vitro and in vivo [11]. The improved system requires only 20 nucleotides on the sgRNA to ensure pairing with the target DNA sequence adjacent to a DNA protospacer motif referred to as protospacer adjacent motif (PAM), needed to constrain the endonuclease Cas9 activity to the target site.

The system has been constantly improved; for example, engineered Cas9 with enhanced specificity could be used for therapeutic approaches. The catalytically inactive form of the Cas9, known as death Cas9 or dCas9, can be used to create fusion proteins that can be targeted to specific loci. Such dCas9 can be coupled to activator or repressive proteins that, with the appropriate sgRNAs, can be directed to induce activation or repression of target genes [14, 15]. The dCas9 in conjunction with specific sgRNAs can also functionally modulate the epigenome. The dCas9 can be fused to the DNA methyltransferase Dnmt3a to methylate CpGs in a range of 100 bp from the sgRNA pairing [16, 17]. Alternatively, the dCas9 can be fused to the P300 histone acetyltransferase or the LSD1 histone demethylase inducing the demethylation of the histone H3K4me2 in a range of 350 bp [18, 19]. These are just two examples of a plethora of possibilities in which the dCas9 can act as an epigenome regulator. The dCas9 has also been fused to specific fluorescent reporters that allow for the visualization and identification of genomic regions within the cell nucleus [20–22]. Nowadays, the CRISPR-Cas9 system provides numerous alternatives to genetically and functionally manipulating genomes.

3 CRISPR-Cas9-Mediated Genome Editing Takes Advantage of the DNA Repair Pathway

Genome editing via the CRISPR-Cas9 system takes advantage of endogenous DNA repair pathways activated by the presence of a double-stranded DNA break (DSB). The repair of a DSB in eukaryotic cells can result in a random insertion or deletion (indels) at the site of the DNA break through the nonhomologous end joining (NHEJ) repair pathway. This strategy has been most frequently used to generate knockouts (*see* below). Otherwise, the homology-direct repair (HDR) pathway allows precise interchange of target sequences, depending on the presence of homologous sequences that precisely recombine at the target genomic location [23]. In summary, DNA sequence insertions or deletions by NHEJ mediates incorporation of different indels into the genome [24]. In contrast, specific modifications can be inserted to target loci taking advantage of the HDR pathway.

4 Mouse Genome Editing Using the CRISPR-Cas9 System

For many years researchers invested time and effort into generating genetically modified mice to study developmental processes, or to generate animal models of human diseases. Today, global and conditional knockout, knock-in, large chromosomal deletions, inversions, and specific mutations can be induced in mice and other organisms using the CRISPR-Cas9 system.

The CRISPR-Cas9 and sgRNA(s) components can be introduced by microinjection into one-cell stage mouse embryos to generate offspring carrying targeted genomic modifications. The procedure requires four main steps: (1) design of the guide RNAs; (2) synthesis, isolation, and purification of RNA and DNA fragments to be microinjected; (3) isolation of one-cell-stage mouse embryos, microinjection of the CRISPR-Cas9 components, and transfer of injected embryos into pseudopregnant mice, and (4) genotyping of offspring to identify descendants carrying the edited genomic DNA [25, 26].

The CRISPR-Cas9 components can be introduced into zygotes as mRNA that gets translated into the Cas9 protein. Cas9 mRNA and the sgRNAs can be in vitro-synthesized from a linearized template plasmid. Alternatively, the sgRNAs can be obtained commercially as synthesized oligonucleotides. A more complex alternative is the design and preparation of repair DNA templates like single-stranded DNA with homology arms about 60 base pairs long for the insertion of short sequences. An example of the use of this is the incorporation of a modified transcription factor binding sequence, or *loxP* sites. Otherwise, more complicated vectors can be created with longer homology arms (over 500 bp to 1.5 kb) to generate larger genomic modifications. Simple deletions and insertions could be identified by DNA sequencing. In addition, replacement mutations and insertion of large DNA fragments can be detected by PCR amplification. Sequencing amplification products will then reveal genome editing events.

The level of CRISPR-Cas9 nuclease activity is critical for success when introducing a desired modification in applications where allele-specific modification is required [27]. For example, several research groups are concentrating their efforts to selectively generate monoallelic or biallelic sequence changes with high efficiency and accuracy. For instance, it is useful to generate human induced pluripotent stem cells (iPSCs) with heterozygous and homozygous genetic modifications in an Alzheimer's disease model [27].

5 Practical Use of the CRISPR-Cas9 System

Genetic mosaicism might mask phenotypes. For example, in mice and zebrafish the injection of sgRNAs targeting the *tyrosinase* gene results in embryos showing mosaic pigmentation [28]. For complete phenotype penetration the genome editing must occur in the germ cells. In *Drosophila* mosaicism is avoided by using transgenic flies that selectively express Cas9 in the germ line [29].

One of the most innovative uses of the CRISPR-Cas9 system is the generation of RNA libraries to perform in situ saturating mutagenesis [30]. This strategy has been used to identify functional enhancers in human and mouse. This strategy allowed editing of primary mouse and human progenitor cells to validate the requirement of the *BCL11A* in fetal hemoglobin gene activation and to identify multiple enhancers [30].

Recent genome wide association studies (GWAS) support the fact that a subset of genetic variants associated with complex diseases affect the function of regulatory elements, particularly enhancers [31]. Such genetic variants seem also to affect the epigenetic components associated with the enhancer's mechanism of action. The CRISPR-Cas9 system has been used to systematically dissect the effect of single nucleotide polymorphisms (SNPs) in the function of regulatory elements at the transcriptional and epigenetic level in human iPSCs [32]. For example, genome editing of a variant in an enhancer element that regulates the expression of *α-synuclein*, which is associated with Parkinson's disease, demonstrated the requirement of the transcription factors EMK2 and NKX6–1 for proper gene regulation. This exemplifies the application of CRISPR-Cas9 system as a tool to understand the contribution of known genetic variants to complex diseases.

For decades, therapeutic strategies for hemoglobinopathies, in particular sickle cell disease and β-thalassemia, have been intensively studied. Using the CRISPR-Cas9 system in combination with adeno-associated viral vector delivery researchers induced homologous recombination in the *HBB* gene in hematopoietic stem cells to correct the pathologic phenotype [33]. This supports the viability of genome editing ex vivo to correct mutations in the β-globin *HBB* gene in patient-derived hematopoietic stem cells to subsequently transplant them for therapeutic purposes.

Another example of a disease that might be corrected using the CRISPR-Cas9 system is Duchenne muscular dystrophy (DMD). A recent report designed different editing strategies using CRISPR-Cas9 to correct a mutation in the dystrophic *mdx*^{4cv}mouse [34]. After correcting the mutation, over 70% of muscle cells expressed normal levels of dystrophin, resulting in increased force generation after intramuscular delivery.

In *Drosophila melanogaster* the use of the CRISPR-Cas9 system has been useful in developmental studies. The CRISPR-Cas9 method has recently been used to perturb the neurogenic gene networks uncoupling EGF synthesis in mutant embryos with a targeted deletion causing loss of function of *rhomboid* neurogenic ectoderm enhancers [35]. This mutation affected expression of intermediate neuroblasts defective (*ind*) leading to defects in central nervous system patterning [35].

6 Genome Editing to Understand the Function of Chromatin Domains

The conformation of the genome into chromatin contributes to differential gene expression during cell determination and organism development. Promoter-enhancer communication is required for coordinated gene expression in time and space. In addition, chromatin domains including a gene or group of genes are organized in the three-dimensional space inside the cell nucleus, creating the so-called compartments A and B, that are further classified as Topologically Associating Domains or TADs [36, 37]. Increasing evidence suggests that alteration of chromatin domains cause genetic modifications with can have serious consequences on developmental processes.

The CRISPR-Cas9 genome editing system has been of great utility to demonstrate that chromatin domain perturbations can cause gene expression alteration in key loci. This is the case for the chromatin domains associated to genes related to T cell acute lymphoblastic leukemia (T-ALL). 40 of 55 genes implicated in the T-ALL are located in a well-defined chromatin domain in Jurkat cells [38]. Among these insulated genes, Young and colleagues identified active oncogenes and inactive proto-oncogenes. Interestingly, CTCF participates in the isolation of these domains, particularly in silent proto-oncogenes [38]. Among many genes, the authors focused in the *TAL1* locus that contains the *TAL1* gene, which encodes a transcription factor overexpressed in most T-ALL cases acting as an oncogene [38]. Then, using the CRISPR-Cas9 system they deleted a TAL1 domain boundary in human embryonic kidney cells (HEK-293T cells). Importantly, in these cells the *TAL1* locus shows a repressive chromatin configuration consistent with the low levels of *TAL1* gene expression. However, a CRISPR-Cas9 mediated deletion of the boundary region containing a binding site for CTCF cause 2.3-fold induction of the *TAL1* transcript [38]. Authors proposed a model in which the integrity of the topological organization of the TAL1 domain is required for the silent state of the *TAL1* proto-oncogene. Similar results were obtained with the LMO2 locus that includes the *LMO2* gene, which encodes for another transcription factor also over expressed and that adopts oncogenic properties in T-ALL.

In conclusion, disruption of a chromatin domain boundary associated with CTCF binding, either by naturally occurring deletion, or induced by CRISPR-Cas system, leads to oncogene activation in cancer cells.

The CRISPR-Cas system has also been employed to manipulate topologically associated domains in the context of syndromes [36, 39]. It is known that human limb malformations are originated by genomic deletions, inversions and duplications. A detailed study has shown that TAD organization can be perturbed, affecting the function of long distance regulatory elements. To further understand the mechanisms linked to TAD organization of these loci in humans, genome editing was performed to generate genetically modified mice with the corresponding genomic rearrangements. Once again, these rearrangements are linked to the specific disruption of CTCF binding sites [39]. One of the most relevant observations is that disruption of TADs causes aberrant chromatin domain configuration affecting gene expression. Another conclusion derived from CRISPR-Cas manipulation in mice is the importance of the CTCF-dependent boundaries in the configuration and integrity of the TADs. Diseases can originate by structural disruption of higher-order chromatin organization and therefore, those human disease-associated deletions result from ectopic enhancer-promoter contacts causing abnormal gene expression. It is important to outline that such 3D genomic structure change is not only due to genomic deletions at TAD boundaries, since duplications and inversions can also cause pathological phenotypes. This study reveals the flexibility of the CRISPR-Cas9 genome editing system, and its usefulness to understand human pathologies and developmental processes.

7 Genome Editing in Embryonic and Induced Pluripotent Stem Cells

Investigations on embryonic stem cells (ESCs) and more recently, human induced pluripotent stem cells (iPSCs) have contributed to better understanding of the fundamental mechanisms of cell proliferation and differentiation. The advent of iPSC technology and the CRISPR-Cas9 genome editing system, fueled investigation on therapeutic strategies owing to the possibility to manipulate, correct or screen for genetic defects. These technologies have facilitated reprogramming of patient somatic cells into iPSCs by controlled overexpression of factors initially found to be enriched in stem cells Oct4, Soc2, Klf4, Myc, and Nanog. iPSCs can now be modified by the CRISPR-Cas system and then, induced to differentiate into somatic cells that could be used as models of human diseases, or in transplantation for experimental therapy [40, 41].

Many different diseases could potentially be targeted by CRISPR-Cas technology. One example is sporadic diseases that carry multiple risk alleles identified by GWASs that can affect chromatin structure in regulatory elements including chromatin boundaries and enhancers [42]. Therefore, an interesting possibility is to generate iPSCs from patients edited by CRISPR-Cas9 to create isogenic cells carrying corrected disease risk variants to enable deciphering the mechanisms linked to polygenic disease.

Many cognitive diseases are associated with genetic defects. The Angelman syndrome is caused by deletion of the maternally inherited *UBE3A* allele, with a phenotype associated with intellectual disability, ataxia and developmental delay. Taking advantage of the CRISPR-Cas system the Angelman syndrome was modeled in cells by knocking out the UBE3A allele [43]. Such modification showed changes in action potential firing and altered synaptic activity. Therefore, the use of genome editing recapitulates cellular phenotypes that allow the investigation of the mechanisms underlying human syndroms, and to test novel pharmacological and therapeutic strategies.

Thus, the CRISPR-Cas9 system has enormous potential to impact on regenerative medicine and disease modeling.

8 CRISPR-Cas9 System for Human Therapy

Along with the discovery of somatic and germ line mutations responsible for specific disease phenotypes, gene therapy arose as a promising alternative to study, treat, and prevent a plethora of human diseases. In addition, genome engineering in animal models including mice and nonhuman primates has also been extensively used to model human diseases, and represent an important tool to study relevant human genetic disorders. In the last decades, TALEN and ZFN-based ex vivo genetic modifications of human iPSCs [44] and other patient-derived hematopoietic cell types (vgr. T cells for HIV and cancer treatments) has demonstrated to be an efficient way to reverse or even protect the host from viral infection [45, 46] or increase antitumor responses [47]. As expected, based on its numerous advantages and efficiency, genome editing with TALEN and ZFN for therapeutic purposes is rapidly being replaced by the CRISPR-Cas9 system. In 2013, the first report using CRISPR-Cas9 to disrupt HIV latent virus in human cell lines was published [48]. Since then, numerous studies have focused on genetic engineering using CRISPR-Cas9 to develop models to understand and control infectious diseases, cancer and genetic disorders.

Studies related to infectious diseases include genome-wide screen of CRISPR-Cas9-induced modifications to identify host factors acting as viral targets for HIV (CD4, CCR5, ALCAM, SLC35B2, TPST2) [49], Hepatitis B and C virus (CD81, CLDN1, OCLN, mR-122, CYPA, ELAVL1, RFK, FLAD1) [50, 51], dengue virus (OST complex, TRAP complex, EMC, ERAD) [50], West Nile virus (SPCS1, SPCS3, EMC, OTS complex, TRAP complex, SEL1L, HRD1) [52, 53], Zika virus (EMC, AXL, OST complex TRAP complex) [54], and murine norovirus (CD300LF) [55–57]. Therapeutic strategies to cure herpes and HIV infection in patient derived iPCS using CRISPR-Cas9 have also been reported [58, 59].

Alternatively, gene editing strategies for cancer are not limited to correct or insert mutations to model and mimic malignancies, but also include immunotherapies based on chimeric antigen receptor (CAR) T-cell editing for adoptive transfer. In murine models, CRIPSR-Cas9 mediated somatic insertion of a chromosomal inversion (Inv2 p21p23) present in a subset of human non-small cell lung cancers, was enough to recapitulate lung cancer, and represents the proof of concept to model human cancers using this technology [60]. Similar approaches to induce somatic mutations with CRISPR-Cas9 in mice have been used to model hepatocellular carcinoma, lung, brain, pancreas and breast cancer [61–67]. CAR T cell editing with CRISPR-Cas9 has been more challenging since it demands optimization for knock-in of large DNA fragments into T cells, however a recent report successfully achieved this and proved that genetically engineered T cells have enhanced antitumor activity in vivo [68].

Examples of the use of the CRISPR-Cas9 technology in the study of other genetic disorders includes murine models to correct DMD (an X-linked fatal disease described above) in murine germ line and stromal cells [69–72], as well as germ line correction of a dominant mutation on the *Crygc* gene causing cataracts [73], and a Fah mutation corrected in murine hepatocytes in a model of hereditary tyrosinemia [74]. A chromosomal inversion at the *F8* gene was corrected using CRISPR-Cas9 in iPSCs derived from hemophilia patients [75]. Similarly, *CFTR* gene mutations have been corrected in cultured intestinal stem cells from cystic fibrotic patients [76]. Altogether the promising results obtained with CRISPR-Cas9 genome editing tools, suggest potential use to correct a variety of other genetic disorders including diabetes and schizophrenia [77, 78].

9 Conclusions and Prospects

The CRISPR-Cas genome editing system can be considered a revolutionary technology that can be applied to dissect developmental and cellular mechanisms, and to generate disease models in a wide variety of organisms and cultured cells. It is clear that improving the efficiency and specificity of the HDR pathway to integrate specific changes, mutations, and other genome modifications will broaden the applications of the CRISPR-Cas system. Importantly, being able to easily modify specific genomic regions will revolutionize the study of the contribution of regulatory elements in processes like cell proliferation, differentiation, morphogenesis, chromatin dynamics, and disease.

A relevant aspect to be considered is the possibility of off-target Cas9 activity [79]. The Cas9 DNA binding is critical for its multiple applications and requires to be guided by RNA to target sites. Researchers have made several efforts to improve PAM motif requirements [23]. Recent efforts are now focused on generating enhanced versions of the *Streptococcus pyogenes* Cas9 to create variants that reduce substantially the off-target activity of the CRISPR-Cas9 system thus increasing genome editing specificity [80]. An appealing option derives from an alternative CRISPR-Cas system that incorporated the Cpf1 protein, which is a highly efficient single RNA-guided endonuclease [81].

Besides reducing off target effects and cytotoxicity, optimizing delivery according to the cell-type and cell-cycle status (Viral vs nonviral approaches, alternative Cas9); choosing the correct DNA-repair pathway (HDR vs NHEJ, PITCh (precise Integration into Target Chromosome) *vs* HITI (Homology Independent Targeted integration)) will be important to ensure efficient genome editing with the CRISPR-Cas9 system [82, 83], which will be critical in clinical applications.

Modifying human and nonhuman genomes, especially for therapeutic purposes, raises many scientific, medical, legal, and ethical concerns that need to be considered. Even if we reach the highest standards to ensure the safe use of CRISPR-Cas technology, it is yet unclear how it could be translated into treatment. Baltimore and colleagues have recently proposed to implement appropriate and standardized benchmarking methods, organize discussion forums including scientific and bioethics topics, and encourage transparent research [84]. In conclusion, ethical and regulatory guidelines are needed to ensure a healthy equilibrium between the enormous potential of this genome editing technology for the benefit of mankind, and minimizing potential risks.

Acknowledgments

This work was supported by the DGAPA-PAPIIT, UNAM (IN209403, IN203811, and IN201114), CONACyT (42653-Q, 128464, and 220503) and Fronteras de la Ciencia-2015 (Grant 290) to FR-T. DGAPA-PAPIIT, UNAM (IA202116) and CONACyT (CB-2015-01-255287, S0008-2015-2-261227, and INFR-2017-01-280464) to PL-L. RGA-M is doctoral student from Programa de Doctorado en Ciencias Biomédicas, Universidad Nacional Autónoma de México (UNAM) and is recipient of the fellowship 317534 and 25590 (Fronteras) from CONACyT.

References

1. Cohen SN, Chang AC, Boyer HW, Helling RB (1973) Construction of biologically functional bacterial plasmids *in vitro*. Proc Natl Acad Sci U S A 70:3240–3244
2. Jaenisch R, Mintz B (1974) Simina virus 40 DNA sequences in DNA of healthy adult mice derived from preimplantation blastocysts injected with viral DNA. Proc Natl Acad Sci U S A 71:1250–1254
3. Recillas-Targa F, Valadez-Graham V, Farrell CM (2004) Prospects and implications of using chromatin insulators in gene therapy and transgenesis. BioEssays 26:796–807
4. Wolffe E (2016) Corporate profile: Sangamo BioSciences, Inc. Regen Med 11:375–379
5. Urnov FD, Rebar EJ, Holmes MC, Zhang HS, Gregory PD (2010) Genome editing with engineered zinc finger nucleases. Nat Rev Genet 11:636–646
6. Geurts AM, Cost GJ, Freyvert Y, Zeitler B, Miller JC et al (2009) Knockout rats via embryo microinjection of zinc-finger nucleases. Science 325:433
7. Miller JC, Tan S, Qiao G, Barlow KA, Wang J et al (2011) A TALE nuclease architecture for efficient genome editing. Nat Biotechnol 29:143–148
8. Christian M, Cermak T, Doyle EL, Schmidt C, Zhang F, Hummel A, Bogdanove AJ, Voytas DF (2010) Targeting DNA double-stranded breaks with TAL effector nucleases. Genetics 186:757–761
9. Yang H, Wang H, Shivalila CS, Cheng AW, Shi L, Jaenisch R (2013) One-step generation of mice carrying reporter and conditional alleles by CRISPR/Cas-mediated genome engineering. Cell 154:1370–1379
10. Mojica FJM, Montoliu L (2016) On the origin of CRISPR-Cas technology: from prokaryotes to mammals. Trends Microbiol 24:811–820
11. Jinek M, Chylinski K, Fonfara I, Hauer M, Doudna JA, Charpentier R (2012) A programmable dual-RNA-guided DNA endonuclease in adaptive bacterial immunity. Science 337:816–821
12. Gasiunas G, Barrangou R, Horvathe P, Siksnys V (2012) Cas9-crRNA ribonucleoprotein complex mediates specific DNA cleavage for adaptive immunity in bacteria. Proc Natl Acad Sci U S A 109:E2579–E2586
13. Sapranauska R, Gasiunas G, Fremaux C, Barrangou R, Horvath P, Siksnys V (2011) The *Streptococcus thermophilus* CRISPR/Cas system provides immunity in Escherichia coli. Nucleic Acids Res 39:9275–9282
14. Gilbert LA, Horlbeck MA, Adamson B, Villalta JE, Chen Y et al (2014) Genome-scale CRISPR-mediated control of gene repression and activation. Cell 159:647–661
15. Konermann S, Brigham MD, Trevino AE, Joung J, Abudayyeh OO et al (2015) Genome-scale transcriptional activation by an engineered CRISPR-Cas9 complex. Nature 517:583–588
16. Liu XS, Wu H, Ji X, Stelzer Y, Wu X, Czauderna S, Shu J, Daden D, Young RA, Jaenisch R (2016) Editing DNA methylation in the mammalian genome. Cell 167:233–247
17. McDonald JI, Celik H, Rois LE, Fishberger G, Fowler T, Rees R, Kramer A, Martens A, Edwards JR, Challen GA (2016) Reprogrammable CRISPR/Cas9-based system for inducing site-specific DNA methylation. Biol Open 5:866–874
18. Hilton IB, D'Ippolito AM, Vockley CM, Thakore PI, Crawford GE, Raddy TE, Gersbach CA (2015) Epigenome editing by a CRISPR-Cas9-based acetyltransferase activates genes from promoters and enhancers. Nat Biotechnol 33:510–517
19. Kearns NA, Pham H, Tbak B, Genga RM, Silverstein NJ, Garber M, Maehr R (2015) Functional annotation of native enhancers with a Cas9-histone demethylase fusion. Nat Methods 12:401–403
20. Chen B, Gilbert LA, Cimini BA, Schnitzbauer J, Zhang W et al (2013) Dynamic imaging of genomic loci in living human cells by an optimized CRISPR/Cas system. Cell 155:1479–1491
21. Tanenbaum ME, Gilbert LA, Qi LS, Weissman JS, Vale RD (2014) A protein-tagging system for signal amplification in gene expression and fluorescence imaging. Cell 159:635–646
22. Dreissig S, Schiml S, Schindele P, Weiss O, Rutten T, Schubert V, Galdilin E, Mette MF, Puchta H, Houden A (2017) Live cell CRISPR-imaging in plants reveals dynamic telomere movements. Plan J 91(4):565–573
23. Komor AC, Badran AH, Liu DR (2017) CRISPR-based technologies for the manipulation of eukaryotic genomes. Cell 168:20–36
24. Harrison MM, Jenkins BV, O'Connor-Giles KM, Wildonger J (2014) A CRISPR view of development. Genes Dev 28:1859–1872
25. Harms DW, Quadros RM, Seruggia D, Ohtsuka M, Takahashi G, Montoliu L, Gurumurthy CB (2014) Mouse genome editing using the CRISPR/Cas system. Curr Prot Hum Genet 15(7):1–15.7.27

26. Seruggia D, Montoliu L (2014) The new CRISPR-Cas system: RNA-guided genome engineering to efficiently produce any desired genetic alterations in animals. Transgenic Res 23:707–716
27. Paquet D, Kwart D, Chen A, Sproul A, Jacob S et al (2016) Efficient introduction of specific homozygous and heterozygous mutations using CRISPR/Cas9. Nature 533:125–129
28. Yen ST, Zhang M, Deng JM, Usman SJ, Smith CN, Parker-Thornburg J, Swinton PG, Martin JF, Behringer RR (2014) Somatic mosaicism and allele complexity induced by CRISPR/Cas9 RNA injections in mouse zygotes. Dev Biol 393:3–9
29. Ren X, Sun J, Housden BE, Hu Y, Roesel C, Lin S, Liu LP, Yang Z, Mao D, Sun L et al (2013) Optimized gene editing technology for *Drosophila melanogaster* using germ lines-specific Cas9. Proc Natl Acad Sci U S A 110:13904–13909
30. Canver MC, Smith EC, Sher F, Pinello L, Sanajan NE et al (2015) *BCL11A* enhancer dissection by Cas9-mediates *in situ* saturating mutagenesis. Nature 527:192–197
31. Lu Q, Powles RL, Wang Q, He BJ, Zhao H (2016) Integrative tissue-specific functional annotations in the human genome provide novel insights on many complex traits and improve signal prioritization in genome wide association studies. PLoS Genet 12:e1005947
32. Soldner F, Stelzer Y, Shivalila CS, Abraham BJ, Latourelle JC et al (2016) Parkinson-associated risk variant in distal enhancer of *α-synuclein* modulates target gene expression. Nature 533:95–99
33. Dever DP, RO BK, Reinisch A, Camarena J, Washington G et al (2016) CRISPR/Cas9 β-globin gene targeting in human haematopoietic stem cells. Nature 539:384–389
34. Bengtsson N, Hall JK, Odom GL, Phelps MP, Andrus CR, Hawkins RD, Houschka SD, Chamberlain JR, Chamberlain JS (2017) Muscle-specific CRISPR/Cas9 dystrophin gene editing ameliorates pathophysiology in a mouse model for Duchenne muscular dystrophy. Nat Commun 8:14454
35. Rogers W, Goyal Y, Yamaya K, Schvartsman SY, Levine MS (2017) Uncoupling neurogenic gene networks in the *Drosophila* embryo. Genes Dev 31:634–638
36. Lupiáñez DG, Spielmann M, Mundlos S (2016) Breaking TADs: how alterations of chromatin domains results in disease. Trends Genet 32:225–237
37. Bonev B, Cavalli G (2016) Organization and function of the 3D genome. Nat Rev Genet 17:661–678
38. Hnisz D, Weintraub AS, Day DS, Valton A-L, Bak RO et al (2016) Activation of proto-oncogenes by disruption of chromosome neighborhoods. Science 351:1454–1458
39. Lupiánez DG, Kraft K, Heindrich V, Krawitz P, Broncati F et al (2015) Disruptions of topological chromatin domains cause pathogenic rewiring of gene-enhancer interactions. Cell 161:1012–1025
40. Takahashi K, Yamanaka S (2006) Induction of pluripotent stem cells from mouse embryonic and adult fibroblast cultures by defined factors. Cell 126:663–676
41. Hockermeyer D, Jaenisch R (2016) Induced pluripotent stem cells meet genome editing. Cell 18:573–586
42. Merkle FT, Eggan K (2013) Modeling human disease with pluripotent stem cells: from genome association to function. Cell Stem Cell 12:656–668
43. Fink JJ, Robinson TM, Germain ND, Sirois CL, Bolduc KA, Ward AJ, Rigo F, Chamberlain SJ, Levines ES (2017) Disrupted neuronal maturation in Angelman syndrome-derived induced pluripotent stem cells. Nat Commun 8:15038
44. Smith C, Abalde-Atristain L, He C, Brodsky BR, Braunstein EM et al (2015) Efficient and allele-specific genome editing of disease loci in human iPSCs. Mol Ther 23:570–577
45. Tebas P, Stein D, Tang WW, Frank I, Wang SQ et al (2014) Gene editing of CCR5 in autologous CD4 T cells of persons infected with HIV. N Engl J Med 370:901–910
46. Mussolino C, Alzubi J, Fine EJ, Morbitzer R, Cradick TJ et al (2014) TALENs facilitate targeted genome editing in human cells with high specificity and low cytotoxicity. Nucleic Acids Res 42:6762–6773
47. Provasi E, Genovese O, Lombardo A, Magnani Z, Liu PQ et al (2012) Editing T cell specificity towards leukemia by zinc finger nucleases and lentiviral transfer. Nat Med 18:807–815
48. Ebina H, Misawa N, Kanemura Y, Koyanagi Y (2013) Harnessing the CRISPR/Cas9 system to disrupt latent HIV-1 provirus. Sci Rep 3:2510
49. Park RJ, Wang T, Koundakjian D, Hultquist JF, Lamothe-Molina P et al (2017) A genome-wide CRISPR screen identifies a restricted set of HIV host dependency factors. Nat Genet 49:193–203
50. Marceau CD, Puschnik AS, Majzoub K, Ooi YS, Brewer SM et al (2016) Genetic dissection of Flaviviridae host factors through genome-scale CRISPR screens. Nature 535:159–163
51. Moyo B, Bloom K, Scott T, Ely A, Arbuthnot P (2017) Advances with CRISPR/Cas-mediated

gene editing to treat infections with hepatitis B virus. Virus Res pii:S0168–1702(16)30733-X

52. Ma H, Dang Y, Wu Y, Jia G, Anaya E et al (2015) A CRISPR-based screen identifies genes essential for West-Nile-virus-induced cell death. Cell Rep 12:673–683
53. Zhang R, Miner JJ, Gorman MJ, Rausch K, Ramage H (2016) A CRISPR screen defines a signal peptide processing pathway required by flaviviruses. Nature 535:164–168
54. Savidis G, McDougall WM, Meraner P, Perreira JM, Portmann JM et al (2016) Identification of zika virus and dengue virus dependency factors using functional genomics. Cell Rep 16:232–246
55. Orchard RC, Wilen CB, Doench JG, Baldridge MT, McCune BT et al (2016) Discovery of a proteinaceous cellular receptor for a norovirus. Science 353:933–936
56. Haga K, Fujimoto A, Takai-Todaka R, Miki M, Doan YH et al (2016) Functional receptor molecules CD300lf and CD300ld within the CD300 family enable murine noroviruses to infect cells. Proc Natl Acad Sci U S A 113:E6248–E6255
57. Puschnik AS, Majzoub K, Ooi YS, Carette JE (2017) A CRISPR toolbox to study virus-host interactions. Nat Rev Microbiol 15:351–364
58. Wang J, Quake SR (2014) RNA-guided endonuclease provides a therapeutic strategy to cure latent herpesviridae infection. Proc Natl Acad Sci U S A 111:13157–13162
59. Ye L, Wang J, Beyer AI, Teque F, Cradick TJ et al (2014) Seamless modification of wild-type induced pluripotent stem cells to the natural CCR5Δ32 mutation confers resistance to HIV infection. Proc Natl Acad Sci U S A 111:9591–9596
60. Maddalo D, Manchado E, Concepcion CP, Bonetti C, Vidigal JA et al (2014) *In vivo* engineering of oncogenic chromosomal rearrangements with the CRISPR/Cas9 system. Nature 516:423–427
61. Xue W, Chen S, Yin H, Tammela T, Papagiannakopoulos T et al (2014) CRISPR-mediated direct mutation of cancer genes in the mouse liver. Nature 514:380–384
62. Weber J, Öllinger R, Friedrich M, Ehmer U, Barenboim M et al (2015) CRISPR/Cas9 somatic multiplex-mutagenesis for high-throughput functional cancer genomics in mice. Proc Natl Acad Sci U S A 112:13982–13987
63. Platt RJ, Chen S, Zhou Y, Yim MJ, Swiech L et al (2014) CRISPR-Cas9 knockin mice for genome editing and cancer modeling. Cell 159:440–455
64. Sanchez-Rivera FJ, Papagiannakopoulos T, Romero R, Tammela T, Bauer MR et al (2014) Rapid modeling of cooperating genetic events in cancer through somatic genome editing. Nature 516:428–431
65. Zuckermann M, Hovestadt V, Knobbe-Thomsen CB, Zapatka M, Northcott PA et al (2015) Somatic CRISPR/Cas9-mediated tumour suppressor disruption enables versatile brain tumour modeling. Nat Commun 6:7391
66. Chiou SH, Winters IP, Wang J, Naranjo S, Dudgeon C et al (2015) Pancreatic cancer modeling using retrograde viral vector delivery and *in vivo* CRISPR/Cas9-mediated somatic genome editing. Genes Dev 29:1576–1585
67. Annunziato S, Kas SM, Nethe M, Yücel H, Del Bravo J et al (2016) Modeling invasive lobular breast carcinoma by CRISPR/Cas9-mediated somatic genome editing of the mammary gland. Genes Dev 30:1470–1480
68. Eyquem J, Mansilla-Soto J, Glavridis T, van der Stegen SJ, Hamieh M et al (2017) Targeting a CAR to the TRAC locus with CRISPR/Cas9 enhances tumour rejection. Nature 543:113–117
69. Nelson CE, Hakim CH, Ousterout DG, Thakore PI, Moreb EA et al (2016) *In vivo* genome editing improves muscle function in a mouse model of Duchenne muscular dystrophy. Science 351:403–407
70. Tabebordbar M, Zhu K, Cheng JK, Chew WL, Widrick JJ et al (2016) *In vivo* gene editing in dystrophic mouse muscle and muscle stem cells. Science 351:407–411
71. Long C, Amoasii L, Mireault AA, McAnally JR, Li H et al (2016) Postnatal genome editing partially restores dystrophin expression in a mouse model of muscular dystrophy. Science 351:400–403
72. Wu Y, Zhou H, Fan X, Zhang Y, Zhang M et al (2015) Correction of a genetic disease by CRISPR-Cas9-mediated gene editing in mouse spermatogonial stem cells. Cell Res 25:67–79
73. Wu Y, Liang D, Wang Y, Bai M, Tang W et al (2013) Correction of a genetic disease in mouse via use of CRISPR-Cas9. Cell Stem Cell 13:659–662
74. Yin H, Xue W, Chen S, Bogorad RL, Benedetti E et al (2014) Genome editing with Cas9 in adult mice corrects a disease mutation and phenotype. Nat Biotech 32:551–553
75. Park CY, Kim DH, Son JS, Sung JJ, Lee J et al (2015) Functional correction of large factor VIII gene chromosomal inversions in hemophilia A patient-derived iPSCs using CRISPR-Cas9. Cell Stem Cell 17:213–220

76. Schwank G, Koo BK, Sasselli V, Dekkers JF, Heo I et al (2013) Functional repair of CFTR by CRISPR/Cas9 in intestinal stem cell organoids of cystic fibrosis patients. Cell Stem Cell 13:653–658
77. Zhuo C, Hou W, Hu L, Lin C, Chen C et al (2017) Genomic editing of non-coding RNA genes with CRISPR/Cas9 ushers in a potential novel approach to study and treat Schizophrenia. Front Mol Neurosci 10:28
78. Gerace D, Martiniello-Wilks R, Nassif NT, Lal S, Steptoe R et al (2017) CRISPR-targeted genome editing of mesenchymal stem cell-derived therapies for type 1 diabetes: a path to clinical success? Stem Cell Res Ther 8:62
79. Kuscu C, Arslan S, Singh R, Thorpe J, Adli M (2014) Genome-wide analysis reveals characteristics of off-target sites bound by the Cas9 endonuclease. Nat Biotechnol 32:677–683
80. Slaymaker IM, Gao L, Zetsche B, Scott DA, Yan WX, Zhang F (2016) Rationally engineered Cas9 nucleases with improved specificity. Science 351:84–88
81. Zetsche B, Gootenberg JS, Abudayyeh OO, Slaymaker IM, Makarova KS et al (2015) Cpf1 is a single RNA-guided endonuclease of a class 2 CRISPR-Cas system. Cell 163:759–771
82. Nakade S, Tsubota T, Skane Y, Kume S, Sakamoto N et al (2014) Microhomology-mediated end-joining-dependent integration of donor DNA in cells and animals using TALENs and CRISPR/Cas9. Nat Commun 5:5560
83. Susuki K, Tsunekawa Y, Hernandez-Benitez R, Wu J, Zhu J et al (2016) *In vivo* genome editing via CRISPR/Cas9 mediated homology-independent targeted integration. Nature 540:144–149
84. Baltimore D, Berg P, Botchan M, Carroll D, Alta Caro R et al (2015) A prudent path forward for genomic engineering and germline gene modification. Science 348:36–38

Index

Paul Delgado-Olguin (ed.), *Mouse Embryogenesis: Methods and Protocols*, Methods in Molecular Biology, vol. 1752, https://doi.org/10.1007/978-1-4939-7714-7,

Printed in the United States
By Bookmasters